Organic Fruit Growing

ORGANIC FRUIT GROWING

K. Lind, G. Lafer, K. Schloffer, G. Innerhofer and H. Meister

CABI Publishing

CABI Publishing is a division of CAB International

CABI Publishing
CAB International
Wallingford
Oxon OX10 8DE
UK

Tel: +44 (0)1491 832111
Fax: +44 (0)1491 833508
Email: cabi@cabi.org
Web site: www.cabi-publishing.org

CABI Publishing
44 Brattle Street
4th Floor
Cambridge, MA 02138
USA

Tel: +1 617 395 4056
Fax: +1 617 354 6875
Email: cabi-nao@cabi.org

Original edition published in German under the title:
Biologischer Obstbau, Copyright © Leopold Stocker Verlag,
Graz, Austria.

A catalogue record for this book is available from the British
Library, London, UK.

Library of Congress Cataloging-in-Publication Data
Biologischer Obstbau. English
    Organic fruit growing / K. Lind ... [et al.].
        p. cm.
Includes index.
    ISBN 0-85199-640-X (alk. paper)
    1. Fruit-culture. 2. Organic farming. I. Lind, K. (Karl) II. Title.
    SB357.24.B5613 2003
    634--dc21
                                        2003002348

ISBN 0 85199 640 X

Typeset in 10pt Melior by Columns Design Ltd, Reading.
Printed and bound in the UK by Cromwell Press, Trowbridge.

Contents

Authors

Production: **Gottfried Lafer**[1] Andi Schmid[4] Karl Lind[1]
Fabio Bonisolli[2] Hans Josef Weber[5] Karl Waltl[6]
Hans Vijverberg[3] Martin Balmer[5] Franco Weibel[4]

Plant protection: **Karl Schloffer**[1] Martin Balmer[5]
Costanzo Menapace[2] Andi Schmid[4]
Arie Kool[3]

Processing: **Georg Innerhofer**[7]
Rino Minutolo[2]
Peter van Meurs[3]

Marketing: **Hans Meister**[1]
Bruno Dorigati[2]
Guus J. van der Horst[3]

Editor: **Karl Lind**[1]

Translator: This English edition was translated from the German by **Alan Leeson**.

[1]Gleisdorf Horticultural College, Styria, Austria
[2]Agricultural Institute, S. Michele all'Adige, Trento, Italy
[3]Helicon Training College, Tiel, The Netherlands
[4]Organic Farming Research Institute, Frick, Switzerland
[5]State Teaching and Research Centre for Agriculture, Viticulture and Horticulture, Bad Neuenahr, Ahrweiler, Germany
[6]Styrian Chamber of Agriculture and Forestry, Graz, Austria
[7]Haidegg Research Centre for Fruit Growing and Viticulture, Graz, Austria

Preface

Organic fruit growing is in vogue. Not only does it make a valuable contribution to the protection of the environment, in areas with predominantly small farms it also helps to provide jobs in rural communities.

Training (both initial training and further training) is essential for meeting the increasing demand for organically grown fruit and processed products made from it. This development has been recognized by the Commission of the European Communities, which has given financial support to the LEONARDO pilot project, 'Developing and testing a teaching module on ecological (organic) horticulture'.

This book, Organic Fruit Growing, is one of the results of this pilot project. It is intended to be used as a basis for work in schools and colleges. It is also intended as a reference book for growers, helping them to decide whether to switch to organic production, where the promising market opportunities at the present time are, of course, accompanied by higher production risks.

Staff of the following schools and institutions collaborated in the pilot project:

- Agricultural Institute, S. Michele all'Adige, Trento, Italy
- Helicon Training College, Tiel, The Netherlands
- Gleisdorf Horticultural College, Styria, Austria
- Organic Farming Research Institute, Frick, Switzerland
- Haidegg Research Centre for Fruit Growing and Viticulture, Graz, Austria
- Office of the Styrian Provincial Government, Department of Agricultural Education, Graz, Austria.

Without the excellent collaboration between all the people involved in the project, it would have been impossible to produce this book, which represents an attempt to bring together knowledge and experience of organic fruit growing from a number of fruit-growing areas of Europe.

We would like to express our sincere thanks to all who have worked on this book.

Walter Eccli	*Karl Lind*	*Jan Gerrits*
(Project Coordinator – Italy)	(Overall Coordinator – Austria)	(Project Coordinator – The Netherlands)

1 Principles of organic fruit growing

General principles

The cultivation of the land and keeping of livestock is bound to have an impact both on the farmed landscape and the natural environment, and thus on the living space and recreational area used by the entire population. The farmer has an obligation to keep this vital resource healthy and functional in the long term, thus ensuring the basis for a healthy lifestyle. Organic farming attempts to fulfil these requirements to the highest possible standards by avoiding the use of synthetic chemical aids. By farming in accordance with the inherent laws of nature, i.e. maintaining energy and nutrient cycles which are as closed as possible, it aims to cause the least possible disturbance to natural equilibria.

In practical terms this means:

- not using readily soluble mineral fertilizers, but instead making careful use of farm-produced natural fertilizers (compost, aerated slurry), green manure, mulches, varied crop rotations and careful tilling of the soil;
- not using herbicides, but instead mechanical or thermal companion crop control and cover crop management;
- not using synthetic chemical pesticides, but instead increasing soil health, choosing species which are appropriate for the location, using resistant varieties and using natural active agents.

In general terms two directions can be distinguished in organic farming: **bio-dynamic** and **organic-biological**.

Bio-dynamic farming, as developed by Dr Rudolf Steiner, a German philosopher and scientist, was the first movement towards organic farming. Steiner concerned himself with agricultural questions only at a theoretical level. His aim was a self-sufficient form of agriculture which also takes into account all the quintessential forces operating at the most minute level of existence. Cosmic constellations are also taken into account in soil cultivation, weed control, and the making and use of preparations. Insofar as weather conditions allow, the practical application of bio-dynamic farming takes account of cosmic forces in sowing, planting, husbandry and harvesting. In his writings Steiner describes the production and action of certain plant-based preparations (yarrow, camomile, nettle, oak bark, dandelion and valerian) which are used in the production of farm fertilizer. Other preparations are horn manure and horn silica, which support terrestrial or cosmic force flows. Particular attention is paid to preventive plant-strengthening measures. In addition, animals should be kept on the farm.

Organic–biological farming, as propounded by Dr Hans Müller, a Swiss agrarian politician, did not develop until the 1960s. Müller's aim was to make farms as independent as possible of bought-in products. He tried to give farmers more self-confidence by assigning them responsibility for public health. A healthy soil is the essential precondition for healthy plants and animals, and consequently also for healthy human beings. The aim of fertilization is to nourish the soil organisms. Efforts are made to minimize the use of artificial fertilizers and pesticides by the techniques of crop rotation with a wide variety of crops and the creation of favourable conditions for beneficial species.

The ecological balance is often disturbed by external factors, which may originate from nature or from man's activities. An increased incidence of pests can be caused merely by disturbances due to the weather. Every type of farming thus constitutes interference with ecological relationships and results in plant protection problems.

Integrated production can be regarded as an intermediate stage towards organic fruit growing. According to the technical definition, 'integrated plant protection' describes a programme which incorporates all possible methods of plant protection, including chemical methods. To avoid causing additional disturbance to the ecological balance, chemical measures are only applied if the economic damage threshold is exceeded. Unfortunately this economic damage threshold is not always the same as the ecological damage threshold.

Legal aspects

Organic farming is the only method of food production which is governed by a European Union Regulation: '*Council Regulation (EEC) No. 2092/91 on organic production of agricultural products and indications referring thereto on agricultural products and foodstuffs*'. This Regulation covers rules of production, labelling, the inspection system, and imports from third countries.

The rules on labelling are contained in Article 5: '*Labelling*'. The labelling and advertising of an agricultural product may legally refer to organic production methods only where *all* the requirements of this

Regulation are met. The following terms in particular are to be used: 'from organic farming', 'from organic agriculture'.

Products which have not yet been packaged for the final consumer must be provided with a label and must be transported in closed packaging or containers sealed in a manner preventing substitution of the contents. In addition to the usual labelling requirements, the labelling must also contain the following information: the name and address of the person responsible for the production or preparation of the product, the name of the product, and the code number of the inspection body.

What is of greatest importance for agricultural production methods, apart from Article 6, *'Rules of production'*, are the rules contained in the Annexes.

The *'Rules of production'* (Article 6) state *inter alia* that the organic production method implies that for the production of agricultural products the requirements of Annex I, at least, must be satisfied. Only products composed of substances mentioned in Annexes I and II may be used as plant protection products, cleaning products, fertilizers or soil conditioners. Moreover, they may be used only insofar as the corresponding use is authorized in accordance with national provisions. Only seed or vegetative propagating material produced by the organic production method may be used. Up until 31 December 2003, conventionally produced seeds and vegetative propagating material may also be used if it can be shown to the satisfaction of the inspection body that no organically propagated material of the appropriate variety is available.

Annex I sets out the principles of organic production, with regard to plants and plant products at the farm level.

Point 1 deals with the length of the conversion period. The principles set out in the Regulation must have been applied for at least 2 years before sowing or, in the case of perennial crops (fruit growing), at least 3 years before the first harvest. The conversion period can be extended or reduced in certain cases.

Point 2 covers methods for maintaining or increasing the fertility and biological activity of the soil. This includes the cultivation of legumes, green manures or deep-rooting plants, the implementation of appropriate crop rotation and the incorporation of organic material from organically run holdings. Other organic or mineral fertilizers mentioned in Annex II may be applied only as a complement. Plant-based preparations (biodynamic preparations) or microorganisms may be used for composting.

Point 3 lists the permitted measures for controlling pests, diseases and weeds. They include the choice of appropriate species and varieties, with an appropriate rotation programme, and the protection of beneficials by the creation of favourable conditions or the release of natural enemies of pests. Weeds may be controlled only by mechanical or thermal methods.

In **Point 4** even the collection of edible wild plants is regulated.

Annex II is divided into Parts A and B. The substances contained in Part A, '*Fertilizers and soil conditioners*', and in Part B, '*Pesticides*', may be applied as a complement only if the measures listed in Annex I are not adequate or if there is an immediate threat to the crop.

Annex III: Minimum inspection requirements

Any operator who produces or prepares organic products or imports them from a third country, with a view to marketing, must be inspected.

Production must take place in a unit in which the land parcels and production and storage locations are clearly separated from those of any other unit not producing in accordance with the rules laid down in the Regulation. A full description of the unit must be drawn up when the inspection arrangements are first implemented. Each year the producer must notify the inspection body of his schedule of production of crop products, giving a breakdown by parcel. He must also keep written and/or documentary accounts showing the origin, nature and quantities of all raw materials bought, and the use of such materials. In addition, accounts must be kept of the nature, quantities and consignees of all agricultural products sold. As well as unannounced inspection visits, every unit is given a full physical inspection at least once a year. All ingredients, additives, adjuvants and formulations used in processing must be declared. The producer must, for inspection purposes, allow access to the storage and production premises and to the parcels of land, as well as to the accounts. He must also provide the inspection body with any information deemed necessary. A product advertised as organic must be clearly declared and recognizable as such at all places and production premises where it occurs physically or in the form of invoices, delivery notes, etc. A key requirement is the traceability of the flow of goods from the producer via processing and trade to the consumer. Only in this way can the creditability of the products, and thus of organic production, be maintained.

Producers who only resell packaged goods are exempt from inspection.

Annex VI lists ingredients of agricultural and non-agricultural origin and permitted processing aids.

Labelling

A few further passages from Regulation 2092/91 are quoted below:

> Even if the use of synthetic plant protection products is required by the authorities (compulsory phytosanitary measures, e.g. in cases of fire blight), the product shall not be labelled as organically produced.

> Plants of the same variety may not produced in production units which are run by the same owner, both organically and conventionally. [Perennial crops (fruit growing and viticulture) which are gradually converted from conventional to organic agriculture are exempt from this prohibition.]

Products may under certain circumstances bear indications regarding their status during the conversion period: *'product under conversion to organic farming'*. The requirements for this are: a conversion period of at least 12 months before the harvest; the product contains only one crop ingredient of agricultural origin; and the name or code number of the inspection body is stated.

Conversion of a conventionally run production unit to organic fruit growing

Fruit-growing areas which still have an intact ecological balance and are thus immediately suitable for organic production exist only in the form of old low-density orchards (scattered fruit trees on farmland), without any fertilization or plant protection measures. It is essential, therefore, to change hitherto conventionally run intensive production units via the roundabout route of **integrated production** and the **conversion phase** into organically managed units. This process must be carried out in stages, since the use of artificial aids (plant protection agents and fertilizers) is very severely restricted, and attention must be paid to the following points:

- improving the ecological balance in the unit
- a positive commitment of growers to the organic production method
- professional training of the grower
- minimizing economic losses in the event of a temporary crop failure.

In order to minimize losses, it may be wise to combine organic production with the production of processed fruit products from fruit obtained from this organic production (see Chapter 6 on *Fruit processing*). This makes it possible to offer a wide range of products which can also be sold by direct marketing (see Chapter 7 on *Marketing*). In this way fruit which cannot be sold on the fresh fruit market because of slight quality defects can still be processed as a value-added product. Fruit processing and direct marketing, however, are highly labour-intensive enterprises which require corresponding resources from the producer, either in the form of technical facilities or in the form of spare labour capacity.

In the conversion phase, only parts of the production unit are managed on an organic basis in any given year, so as to gain experience of this method of production and keep economic losses within acceptable limits. Complete conversion of an establishment takes 3–4 years on average (the minimum being 36 months). In some cases the conversion process can take considerably longer. During this conversion phase the fruit cannot yet be sold as organic fruit, but only as a 'product under conversion'. The prices of products of the latter type are hardly any higher than those of fruit from integrated production, so lost revenue must be expected during the conversion period, not just because of low prices but also because of falls in yield and quality.

In this **transitional phase** (integrated production) it is advisable:

- to implement production on integrated lines
- to carry out measures for the improvement of soil structure
- to take soil remediation measures (e.g. humus enrichment, improvement of soil activity)
- to use organic fertilizers, which allow soil life to be regenerated in its broad diversity
- to optimize environmental conditions, with hedges, wild-flower meadows and other ecological elements to encourage beneficials
- working with nature requires a high level of expertise and therefore necessitates the continuous training of fruit growers.

The organically run production unit is based on an ideal combination of a variety of ecological elements:

- the tramlines are full of wild flowers and are mowed alternately
- the rows are not kept clear the whole year round
- there are wild-flower strips at the edges of the plots
- unproductive areas are not mulched, but mown as extensive meadows
- pioneer plants grow on areas not used for fruit growing (e.g. paths and path edges, turning areas)
- hedges and copses with a wide diversity of botanical species provide food and shelter (including protection from the wind) for animals
- individual trees and orchards with standard trees are retained
- individual species of animals are deliberately encouraged with artificial sanctuaries (e.g. nesting boxes, heaps of stones or branches, perches)
- wild-flower patches are created on areas to be replanted.

Checklist for would-be organic fruit growers

Attitude of the grower to organic fruit growing
(positive **+1**, neutral **0**, negative **–1**)   ....

Marketing structure of the establishment
(high percentage of direct marketing, close to market, fruit processing **+1**,
good ratio between direct marketing and sales via wholesalers **0**,
location far from market, unfavourable marketing structures, no possibility
of processing **–1**)   ....

Professional qualifications of the grower
(grower running a first-class establishment with consistent implementation of
integrated production **+1**, grower running a good establishment with high
professional qualifications **0**, environmental fundamentalist, drop-out without
professional qualifications **–1**)   ....

Continued

Checklist for would-be organic fruit growers *Continued*

Previous method of running the establishment
(organic **+1**, integrated **0**, conventional **–1**) ....

Location of the fruit-growing areas
(sheltered, sunny slopes with good drainage, with no risk of frost **+1**,
good gentle slopes, open locations **0**,
closed, low locations with a high risk of frost **–1**) ....

Condition of the soil on which the fruit is grown
(productive soil with good root penetration and optimum humus content **+1**,
normal soil with good nutrient levels **0**, soil with tendency to compaction
and waterlogging, level soil with low humus content **–1**) ....

Range of varieties in the establishment
(partially resistant varieties suitable for organic growing **+1**, normal range **0**,
high proportion of varieties susceptible to disease and pests **–1**) ....

Recycling (composting, mulching, manure, etc.) (is a key element in the
running of the establishment **+1**, is used to some extent in the establishment **0**,
is not used in the establishment **–1**) ....

Total ....

If the total score is more than 4, the grower probably meets the professional and
commercial requirements for conversion to organic production.

2 Planning and setting up an organic production unit

Choice of site: ecological principles

The term **site** encompasses all the factors associated with **climate**, **location** and **soil**. These three factors are closely related to each other. This interaction produces the living conditions for the tree. In addition, there are also interactions of the fruit plants with each other, and interactions between the fruit plants and the relevant fauna and flora (**biocoenosis**).

location
risk of frost
sunshine
altitude
wind conditions

climate
temperature
rainfall
light

soil
soil type
soil structure
pH
humus content
water and nutrient supply
biological activity

Fig. 2.1. Factors determining yield and quality in fruit growing.

The economics of organic fruit production is largely dependent on the site. Since in organic fruit production it is only possible to make very limited use of chemicals for controlling disease and pests, the choice of site is of critical importance as a preventative measure for plant protection. The different requirements of particular species and varieties of fruit call for specific locations. To achieve economic success, the site conditions must match the site requirements of the various species and varieties of fruit as closely as possible. The most important prerequisite for successful organic fruit production is therefore to choose varieties that are right for the location (planting the right variety in a location which is optimal for it). It would be pointless to try to set up profitable organic fruit production in areas where the site conditions are unsuitable. If the site deficiencies are slight, it is appropriate to consider whether it is feasible and cost-effective to remedy them. Before setting up an organic production unit, it is absolutely essential to consult the local advisory service for fruit growers, which knows the special characteristics of the area in question. In this way errors in planning can be avoided.

Climate

Climate is of fundamental importance in determining whether a particular species of fruit can be successfully grown at all in a particular area.

The general climate is shaped by the following factors:

- temperature
- rainfall
- light.

Meteorological stations measure and record the averages and provide a basis for climatic assessment. The climatic conditions of a particular area can be deduced from the general climatic data. These data are summarized in climatic charts, which are a key element in assessing the suitability of an area for fruit production.

Temperature

In order to assess the suitability of a particular area for fruit production, apart from the mean annual temperature it is important to know the mean temperature during the months from May to September (principal growing season).

The mean annual temperature is not in itself a reliable indicator of suitability for fruit production. A better indicator is the length of the growing season, i.e. the number of days of growth: days with a mean diurnal temperature higher than +5°C (physiological zero). The length of the growing season in a particular area is determined by the number of days of growth, and for successful organic fruit production it should be more than 235 days. In central Europe, the temperature boundary for intensive production of pome fruits lies roughly along the line where, as a long-term average, apple trees blossom before 15–20 May. If blossoming starts later, many varieties of pome fruit (except for early varieties) cannot be guaranteed to ripen fully.

Figure 2.2 shows the differences in mean annual temperature in different fruit-growing areas (Styria, Austria; Trentino, Italy; and The Netherlands).

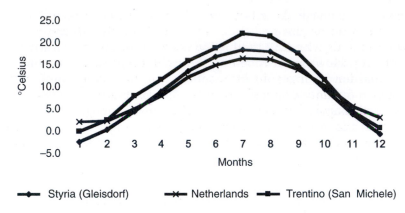

Fig. 2.2. Variation in temperature in different fruit-growing areas.

Table 2.1. Heat requirements of different species of fruit.

High	Low
apricot	some varieties of plums (e.g. Hauszwetschke)
peach	pear (less demanding varieties such as Précoce de
sweet chestnut	Trèvoux and Conference)
pear cultivars	many apple varieties
plum cultivars	sweet cherries
walnut	small fruit
hazelnut	rowan cultivars
quince	

Table 2.2. Heat requirements of different apple varieties.

High	Medium	Low
Idared	Florina	Elstar
Meran	Delbarestivale	Boskoop
	Jonagold	early ripeners (e.g. J. Grieve)
	Pinova	Kronprinz Rudolf
	Fiesta	Topaz
	Elise	Santana

Apples and especially pears show big differences between varieties. Some varieties are very demanding (e.g. winter pear cultivars), others less so (mainly cider fruit varieties).

Winter frost

Severe winters with low extreme temperatures over long periods, without snow cover, can lead to catastrophic frost damage (winter frost damage). The damage caused by low winter temperatures is often underestimated. It occurs mainly at the start and towards the end of the winter. Mild periods with temperatures above zero followed by a sharp temperature drop (down to −15°C) are the most dangerous.

Sudden changes in temperature of this type cause very severe damage to the wood and flower buds of sensitive varieties (e.g. Gala, Elstar, Jonagold, Braeburn) as well as root damage to rootstocks (e.g. quince and M9). It has been found that sunny slopes, hollows and sites with a high groundwater level (moist, cold soil) are most at risk in this respect.

Table 2.3. Sensitivity of different species of fruit to winter frost.

High	Medium	Low
quince	apple	elder
peach	pear	redcurrant
plum cultivars		gooseberry
apricot		
walnut		

Apples and pears show big differences between varieties.

Rainfall

The water requirements of fruit trees are primarily supplied by precipitation in the form of rain and snow.

The most reliable water supply of all can be obtained on deep soils with a high water capacity and with good rainfall distribution during the principal growing season. The amount of water needed by fruit trees increases with increasing mean temperature during the growing season. In dry spells, the usable groundwater level is also important, provided that it is not deeper than 1 m on light soils and not deeper than 2.5 m on heavy soils.

Lack of water causes a reduction in assimilation. In apple trees the following signs are evidence of an acute lack of water:

- limp foliage (loss of turgor)
- limpness and slight drooping of the non-lignified tips of the long shoots.

These symptoms characterize the critical phase of water supply and indicate that the tree definitely needs to be watered. Young trees and newly planted trees are particularly sensitive to drought and need to be watered in critical phases. In order to reduce drought stress, it is also very important to keep the planting area weed-free and/or cover it with various organic materials, such as manure. Care should be taken to ensure that the trees are supplied with enough water in their yield phase, in the period from spring to the beginning of July (additional watering with overhead sprayers, trickle irrigation or micro-sprinklers). Because some of their roots go down deep (except for dwarfing types or cloned rootstocks) and many of the organs are lignified, fruit trees are quite capable of withstanding short periods of drought without damage.

Fig. 2.3. Choice of site: unfavourable locations in shade at the edge of woods, good insolation at the top of the slope; low-lying locations are at more risk of frost than slopes.

Fig. 2.4. South-facing slopes are favourable for organic fruit production.

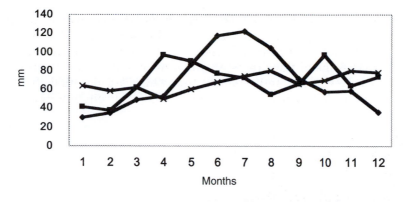

Fig. 2.5. Rainfall pattern of different fruit-growing areas.

Table 2.4. Water requirements of different species of fruit.

High	Low
plum	pear
apple	peach
currant	apricot
elder	walnut
hazelnut	other species of small fruit

Hail

Hail is a form of precipitation which can cause catastrophic damage in some regions – although usually only in a very limited area. In such locations quality fruit production is extremely risky without suitable measures to protect against hail.

Table 2.5. Possibilities of protection against hail.

Active protection against hail	Passive protection against hail
rockets	hail nets
aeroplane	
burner	

Table 2.6. Advantages and disadvantages of the hail net.

Advantages	Disadvantages
protection for fruit and fruiting wood	costs
uniform utilization of storage facilities	additional work during the year
low sorting costs	loss of light (up to about 25%)
guaranteed market presence	poorer colour development
reduction of sunburn damage	leaves wet for longer
less exposure to wind	restriction of mechanization
greater security for the individual grower	

Light and hours of sunshine

The production of quality fruit requires between 1600 and 1800 hours of sunshine per annum. Sunny years have a beneficial effect on both the external and the internal quality of the fruit (firm peel, firm flesh, high sugar content and good flavour quality).

Importance of light for the fruit tree

- Warms up the soil
- assimilation (sugar content of the fruit, maturity of the wood)
- accumulation of yellow pigments in the fruit, breakdown of chlorophyll
- formation of the red surface colour of fruits (anthocyanins)
- development of strong flower buds.

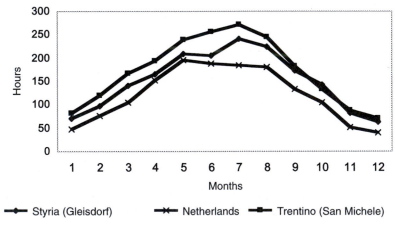

Fig. 2.6. Hours of sunshine in different fruit-growing areas.

There are two periods in which trees are sensitive to lack of light. The first is shortly after flowering (when lack of light results in smaller fruit and can lead to increased fruit drop in June), and the second period is before harvesting (when lack of light affects fruit growth and surface colour).

The insolation available has a marked effect on the yield and quality of the fruit. Irrespective of the density and method of planting, the optimum figure for light utilization is 70%. If because of a too high planting density more than 70% of the light is utilized, i.e. less than 30% of the light reaches the soil, this leads to a drop in quality. Utilization of less than 70% of the light results in good quality but an inadequate yield.

Location

Location means the way the climatic conditions are affected by mountains, hills, towns and villages, forests, altitude, declivity and orientation.

Types of locations

The main types of locations can be classed as open, closed or sheltered.

Open locations

Open locations are exposed in all directions and are therefore very much affected by wind. The advantage that they offer is the low incidence of disease and pests due to better and faster drying. Continuous wind impairs tree growth and insect flight. It also leads to increased wind damage and pre-harvest fruit drop. Marked improvements in this respect can be achieved with wind shelters. Hedges planted as wind shelters can also serve as ecological niches for a large number of beneficials in organic fruit production.

Closed locations

Closed locations are enclosed on all sides by woods, houses, hills or mountains, etc. The lack of air movement leads to an increased incidence of disease and pests. These locations are also subject to a severe risk of frost because they trap cold air.

Sheltered locations

Sheltered locations are the best type of location for fruit growing. They are protected on the north-east side (by woods, buildings, etc.) and are open towards the south-west. This provides the possibility of air flow in the latter direction and reduces the risk of frost.

Criteria for assessment of a site

Frost risk (early and late frost)

Sites prone to frost reduce the reliability of the yield. **Late frosts in spring** can destroy a whole year's crop at a stroke. The damage, however,

Fig. 2.7. A hail protection net and anti-frost irrigation prevent losses of yield and quality.

Fig. 2.8. Anti-frost irrigation prevents damage to blossoms from frost.

Fig. 2.9. Frost rings on fruit.

Fig. 2.10. Hail damage to fruit.

Fig. 2.11. Winter frost damage to the trunk.

does not only consist of the failure of the crop but also in the fact that the frost causes biennial bearing. **Early frosts in autumn** shorten the growing season and can reduce the quality of the crop.

Sites which trap cold air, because of woods, hollows, valleys, groups of buildings, railway embankments, etc., should be avoided. The nature of the terrain must therefore be taken into account, in order to decide whether the site under consideration lies in a cold air pocket. **Wind** is an important factor in this respect. Moderate air movement is beneficial; cold air which is moving seldom leads to frost damage!

Measures to prevent damage from late frosts

Radiation frosts can be reduced in severity or prevented by various measures. The safest method is anti-frost irrigation.

- **Smoke formation**: an artificial fog is created by burning damp organic materials. A temperature rise of around 2–3°C can be achieved with organized smoking.
- **Heating**: oil or gas stoves (150–200 stoves/ha) can be used for heating, thus reducing the effect of frost.
- **Air circulation (wind generation)**: big propellers are used to blow warm air on to the crops; in this way cold air is mixed with warm air.
- **Anti-frost irrigation**: the heat of solidification of water is used to prevent frost damage (335 kJ/litre of water). It is important to switch on at the right time, to make sure that the plants do not get too cold.

Insolation

Only sites which receive good, prolonged insolation should be considered for organic fruit growing. In addition, care should be taken to make sure that there is not too much shade from the edges of woods, higher ground, buildings, etc. In sloping and hilly sites, even slight differences in the direction the slope faces (orientation) can significantly change the length of insolation. **North-facing slopes** should in most cases be regarded as unfavourable for growing fruit (with the exception of apricots); gentle **south-facing slopes** can normally be considered optimal, except that there is often a risk of drought in summer, so that watering may be necessary. South-facing sites get about twice as much light as north-east-facing slopes which are in shade. A south-facing location is not necessarily always the best for some species and varieties of fruit, however. The early rise in temperature causes early budburst and thus early production of blossoms, which are then often damaged by frost. Early flowering fruit species, such as apricot, cherries and early-flowering apple varieties (e.g. Braeburn, Jonagold), are particularly at risk.

Apart from the orientation, the **gradient of the slope** is also a key factor determining insolation. If the gradient is 28% or more, profitable fruit production is virtually impossible, since mechanization is no longer possible or can only be achieved with expensive specialized machines.

Gentle slopes with a gradient of 10% or less are the optimum from the point of view of microclimate and labour costs.

Wind conditions

Wind strength and **frequency** are also important. Frequent strong winds have the following negative effects:

- the soil dries out
- there is less dew
- it is more difficult for insects to fly
- vegetative and generative performance is reduced
- chafing on fruit
- increased pre-harvest fruit drop
- broken branches and fallen trees (in gales).

Effective protection can be provided by planting a hedge as a windbreak. A windbreak which is properly set up at the right time results in an earlier start to fruit production, an increase in yield and better fruit quality because of the significant improvement in the microclimate. A wide variety of broad-leaved woody plants, which should be adapted to the area in question (e.g. ash, sycamore, false acacia, black alder, poplar, dogwood, lilac, elder, black cherry, honeysuckle), are suitable for hedging. Windbreaks usually consist of three or four rows, with one or two rows of poplars together with two shrub hedges. Hedges planted as windbreaks also provide shelter for many natural enemies of fruit production pests (mammals, insectivorous birds, predators, and other beneficials). These numerous advantages are accompanied by some disadvantages, however: reduced air circulation, poorer drying, higher incidence of disease and pests, shade and root competition.

Slight air movement is beneficial for many species of fruit, e.g. apple, apricot and cherry, as it dries the trees quickly after rain and thus reduces the incidence of fungal diseases.

Altitude

With increasing altitude, the air temperature decreases, the climate becomes more severe, and the growing season becomes shorter. The size of the fruit also decreases with increasing altitude, but the internal quality (colouring and firmness) and keeping quality of the fruit usually increase. Fruit produced at altitude usually has a brighter colour because of increased UV radiation. An increase of 30 m in altitude delays fruit maturity by about 1 day.

Soil

The soil is the underground living space of the tree and is thus one of the most important factors for sustainable ecological production. A well-aerated, retentive soil of sufficient depth, with good root penetration and an abundance of soil organisms, is essential for ensuring that plant growth is healthy and reliable – especially in years with climatic extremes such as drought or excessive rainfall. Permanent crops such as fruit trees are especially dependent on a deep root volume. In many cases, cultivated soils suffer from compaction, which can seriously impair tree growth. Although the increasing use of machinery in fruit production reduces labour costs, it also affects soil health (see section in Chapter 4 on *Protection of the soil when using machinery*). A healthy soil structure can also be destroyed if insufficient care is taken in soil cultivation or ground levelling work. The consequences are **compaction**, with **waterlogging** and **lack of oxygen**, or a **dead soil** which shows little or no biological activity. Proper soil preparation is therefore one of the most important prerequisites for successful organic fruit production.

A healthy soil should have the following composition and properties:

- **50% solid constituents** (organic – humus, inorganic – minerals)
- **50% pore space** (half filled with water and half with air)
- **intact soil life** (biological activity of the soil).

Soil porosity is optimal when about half of the soil volume consists of pores; over 10% of these should be large pores, more than 15% medium-sized pores, and less than 20% fine pores.

Intact soil life: an important factor in soil fertility

In organic farming, the self-regulation capacity of the soil and the conservation of the environment through the use of nutrient cycles which are as closed as possible, traditionally have a key role to play. Because of the decision not to use quick-acting corrective agents such as synthetic plant protection products and easily soluble fertilizers, the self-regulation capacity of the soil and the conservation of the environment are important prerequisites for the successful operation of organic growing systems.

The role of soil organisms

- Mineralization of organic matter (mineralization of nitrogen and carbon, etc.)
- humification (conversion of organic matter to stable humic substances)
- improvement of nutrient mobilization and uptake by plants (especially through mycorrhiza, root symbiosis fungi)
- nitrogen fixation from the atmosphere, especially by root nodule bacteria
- protection of the plant from soil pathogens – the 'antiphytopathogenic potential' of the soil

Fig. 2.12. Open sites are exposed to wind – it is advisable to plant a hedge as a windbreak.

Fig. 2.13. Earthworm activity – evidence of intact soil life.

Fig. 2.14. A hedge as a windbreak, with a wild-flower-strip.

Fig. 2.15. Poor decomposition of leaves because of low earthworm activity.

- reducing the potential spread of plant diseases, e.g. reducing scab incidence through increased leaf decomposition
- stabilization of soil structure by bacterial mucus.

The **organic matter** in soil has the following important functions:

- storage of water and nutrients in the root penetration zone of the soil
- loosens heavy soils and increases crumb formation
- light soils become more retentive
- better warming of the soil because of the dark colour of humus (longer growing season)
- provides food for soil organisms
- supplies nutrients and makes CO_2 available for assimilation
- acts as a buffer against drought, sudden temperature changes, etc.

Table 2.7. Soil requirements of the most important fruit species.

Species	Soil requirements
apple	free-draining deep loamy soil
pear on quince	warm, sandy loams without a high lime content
pear on seedling	free-draining, warm, retentive soils
sweet cherry	medium-heavy soils, low lime sensitivity
plum	free-draining, otherwise almost any soil type suitable, depending on rootstock
peach	deep soils which warm up easily

Soil preparation

Soil testing

Soil testing is done in two ways: firstly, assessment of the soil **on site**, and secondly, analysis of soil samples **in the laboratory**.

Soil testing on site

When a soil is tested on site, the soil profile is taken. This means a vertical section through the soil, from the surface down to the unchanged primary material of the soil (parent rock). In practice a hole is dug with a vertical wall, which reveals the profile of the soil.

DETERMINING SOIL DEPTH. The parent rock lies at different levels, depending on soil depth:

shallow soils	<30 cm
moderately deep soils	30–70 cm
deep soils	>70 cm

Increasing depth gives the following advantages:

- greater root penetration area
- higher storage capacity for water, air and nutrients.

DETERMINING THE TYPE OF SOIL. Every soil consists of mineral particles of varying sizes. The ratio of the particle sizes present in a soil largely determines the properties of the soil. The granulometric composition (texture) of a soil can be precisely determined by laboratory tests (sieve and sedimentation analysis). Large particles are determined by sieving and finer ones by sedimentation. The soil type is apparent from the ratio of the individual granulometric fractions.

Table 2.8. Particle size classes of soil (granulometric fractions of fine soil).

Particle size class	Particle size (mm)
sand	2.0–0.06
silt	0.06–0.002
clay	<0.002

On site the soil type is determined by a **finger test**. For this the soil is moistened down to the liquid limit, then kneaded between the fingers and assessed for roughness by rubbing between the fingers, the feel of the individual granules, and deformability (degree and repeatability).

ASSESSMENT OF SOIL STRUCTURE (SOIL TEXTURE). The structure of the soil is important for the water content, air content and thermal metabolism of the soil. It has determinative effects on plant growth and on root penetration, soil life, and the workability of the soil.

Table 2.9. Important forms of soil structure.

Basic structures	Composite structures (aggregate structures)	Soil fragments
single particle structure	plate structure	crumb
overall structure	prismatic structure	clods
	granular structure	
	block structure	
	crumb structure	

THE SPADE TEST FOR ASSESSMENT OF STRUCTURE. It is absolutely essential to test the soil structure with the spade in order to be able to make the correct decisions about soil preparation and to judge the success of soil improvement measures. Two spades are used to take a block of soil at least 10 cm thick. The state of the soil is then described in terms of the following criteria:

- size, structure and distribution of the soil aggregates in the crumb
- quantity and depth of roots
- distribution of capillary roots and nodules
- moisture distribution
- quantity, distribution and degree of decomposition of organic matter worked into the soil (crop residues, manure, etc.).

Assessment of the soil in terms of these criteria gives a very accurate picture of the state of the soil. In particular, the distribution of the crumb structure, roots and moisture very soon reveals where horizons that impede plant growth are present in the soil.

In this assessment the following relationships can be determined between the state of the soil and root development:

- In **crumbly soil** there is abundant, uniform and deep root penetration; the individual roots are extended (crumb <10 mm).
- In **friable soil** (1st-degree compaction) there is less root penetration; the root hairs are undulated and grow around the aggregates (aggregates >10 mm).
- In **soil which breaks into clods** (2nd-degree compaction) there is little root penetration. The roots are usually found only on the fracture surfaces of the clods (clods >50 mm).
- In **soil which breaks into plates** (3rd-degree compaction) there is hardly any root penetration in a vertical direction. The roots are found predominantly in the horizontal fracture surfaces.

TEST FOR EARTHWORM ACTIVITY (KENNEL TEST). The activity of earthworms is estimated using straw. This method can give a rough estimate of the current above-ground activity of species which burrow vertically. Twenty blades of straw, 5 cm in length, are spaced 2.5 cm apart on a defined area (50 × 50 cm), ideally 2–3 days after rain has fallen on fresh soil. The changes in their position are recorded every 24 hours over a number of days. If earthworms have been active during the night this can be seen from the change in the position of individual blades of straw. Faeces on the surface of the soil are also evidence of earthworm activity.

Testing soil in the laboratory

In the laboratory the chemical and physical properties of the soil are tested, as well as nutrients. A soil sample has to be taken beforehand for this purpose and sent to the laboratory.

HOW SHOULD A SOIL SAMPLE BE TAKEN? For accurate soil analysis, it is of critical importance that the soil sample is taken properly. The best way of taking a soil sample is with a soil probe. If a soil probe is not available, a spade can be used instead. The soil removed should be collected in a clean bucket and should be thoroughly mixed. The soil sample should be taken from at least 20–40 different places, uniformly distributed over

Fig. 2.16. Tools for the spade test.   **Fig. 2.17.** Carrying out the spade test.

Fig. 2.18. Subsoil with few roots and dense soil structure – needs to be developed with green manure.

the area under investigation (up to 1 ha). Depending on the laboratory's instructions, the depth at which the sample is taken should be 25–30 cm. Each soil sample must of course be clearly marked, and detailed information about the fruit species, fruit variety, size of the plot, etc., should be entered on the laboratory's data-collection form.

A composite sample of at least 0.25–0.50 kg is needed for testing for individual nutrients. The composite samples should be packed in clean packaging material (paper bags, plastic bags or boxes). Tests on soil for nutrients should be repeated every 3 years.

The soil sample is usually tested for the following parameters:

- **pH** (soil reaction, pH in $CaCl_2$):

optimal values for	pome and stone fruit	6.0–6.5 (7.0)
	small fruit	5.5–6.0
	blueberry and lingonberry	<4.5

- **Humus content**:

minimum humus content for	light soils	1.5%
	moderately heavy soils	2.0%
	heavy soils	2.5%

- **Nutrient analysis**: in order to determine the fertilization needed in organic fruit production, it is important to test the soil for nutrients before planting. This provides information about nutrient deficiencies and can also indicate whether there are excessive levels of certain nutrients in the soil. The soil sample should be tested for the following nutrients before a new orchard is set up:

  phosphate
  potassium
  lime (exchangeable calcium)
  magnesium
  boron
  other trace elements: manganese, zinc, copper, iron.

The results of the soil tests provide information about any fertilization measures that may be required. The nutrient content of the soil is adequate if the following quantities are contained in 100 g of soil:

phosphate (P_2O_5)	11–25 mg/100 g
potassium (K_2O)	11–32 mg/100 g (depending on how heavy the soil is)
magnesium (MgO)	7–13 mg/100 g (depending on how heavy the soil is: Schachtschabel method)
exchangeable calcium	250–300 mg/100 g
boron (B)	0.8 ppm (Baron's method)
manganese (Mn)	70 ppm (EDTA extraction)
zinc (Zn)	8 ppm
copper (Cu)	8 ppm
iron (Fe)	100 ppm.

At low pH, potassium and phosphate are determined by the **DL method** (double lactate), but if the pH is >6.0 they are determined by the **CAL method** (calcium acetate lactate).

- **Measurement of soil activity**: there are a number of laboratory methods which are suitable for measuring the biological activity of the soil. In principle, a distinction is made between **direct** and **indirect methods** for the determination of soil activity. The biomass in the soil, for example, can be estimated by counting the individual organisms in the soil, or the measurement of respiration after the addition of a nutrient in excess can provide an indication of active biomass. Moreover, in determinations of activity, a distinction is made between **actual** and **potential activity**. Actual activity values are values measured at the time that the sample was taken. Determinations of potential activity, on the other hand, show the level of performance that microorganisms are capable of under optimum experimental conditions, after the addition of a nutrient substrate and prolonged incubation.

Basal dressing

Basal dressing is advisable if the results of the soil test indicate that there is an inadequate supply of nutrients in the soil. Only products specified in Annex II of EU Regulation 2092/91 should be considered for use as fertilizers and soil conditioners. If the pH is too low, the soil should be limed with up to 2000–3000 kg of calcium carbonate/ha per year (0.2–0.3 kg/m^2), or with dolomitic lime if there is also magnesium deficiency. If there is not enough humus, it is advisable to enrich the soil with organic matter (see section in Chapter 4 on *Organic fertilizers for soil conditioning*).

Loosening the soil (mechanical preparation of the soil)

In many cases cultivated soils have compacted areas which can severely impair tree growth. Compaction in soil may be due to factors associated with soil formation or geology, or it may have been caused by mechanical pressure from machinery or by deposits by transport and ground-levelling vehicles. If the soil is compacted it is absolutely essential to loosen the subsoil (e.g. by trenching or deep cultivation) before a new orchard is set up. Dwarfing rootstocks for apples and pears (M9, M27, quince C), elder and small fruit trees are particularly sensitive to soil compaction.

Possible ways of loosening the soil

The use of soil tillage equipment before planting depends on the results of the soil tests. If there are layers in the soil which are difficult for the roots of the fruit trees to penetrate, the soil needs to be loosened and then enriched with green manure. The depth of tillage is critical for successful loosening of the soil and should be only slightly (about 5 cm)

Table 2.10. Comparison of compacted and loosened soils.

Compacted soils	Loosened soils
are waterlogged	have a high pore volume
consequently suffer from oxygen deficiency	high gas exchange between soil air and
are therefore cold	atmosphere
humification is inhibited	a lot of O_2 in the soil
putrefaction processes start	the soil warms up more easily
increase in CO_2 in the soil air:	high soil organism activity
inhibits root growth	good root growth
inhibits water and nutrient uptake	
aerobic soil life decreases	
biodegradation ceases	

below the compaction horizon. Soils which are naturally compact (e.g. clay soils, gley soils) can only be mechanically loosened to a limited extent and are therefore less suitable for organic fruit production.

Subsoiling machines

RIGID-TINE CULTIVATOR. The tines have to break up the compacted zone from below. Single-arm cultivators are available from almost all major plough manufacturers. Because of their relatively low purchase price, these machines are the most widely used. Depending on the working depth and size of the shares, single-arm cultivators have power requirements of 50–150 hp (37–110 kW). The tractor drive power is not usually adequate for multiple-arm cultivators, as a minimum share size of 120 × 400 mm is needed to achieve an adequate loosening effect, and the power requirements for two-arm cultivators may be as much as 150 hp (110 kW) when the soil is loosened to a depth of 70 cm. If the shares are too short and too narrow, the soil is not broken up sufficiently from below, but instead the soil all around is compacted. In order to achieve a good lifting effect, the share should be set at an angle of about 25°. The soil needs to be suitably dry if it is to be broken up effectively.

VIBRATING SUBSOILERS. These machines were developed for the purpose of achieving a loosening effect not only through the tractive power but also through the power take-off of the tractor. In this way it is possible to loosen the soil even if the surface does not give a firm hold or the machine has to work uphill. The movement of the shanks to and fro is driven by an eccentric drive.

In some types of vibrating subsoiler only the tine is moved, not the shank and tine together. This needs less power from the power take-off. At a working depth of 80 cm a single-arm subsoiler needs a drive power of at least 80 hp (59 kW) and a two-arm subsoiler needs at least 120 hp (88 kW). Because of their good loosening effect these vibrating subsoilers, which are highly stable, have been found to give good results even in heavy soils.

SPADING MACHINES.   Spading machines work with tools that cut through the soil, and they can easily mix the soil. They can be used down to a depth of 50 cm. A shear bolt safety mechanism prevents the tools being broken if there are unexpected objects in the soil.

MULTIPURPOSE AMELIORATION TECHNIQUES.   With its powered cutting tools the MM 100 subsoiler provides good loosening of the soil down to a depth of 70 cm. The power needed is about 100 hp (73.6 kW) with moderate compaction. A modified version, the MM 50, which can be used with narrow-gauge orchard tractors having a power of 50 hp (37 kW) or more, has been developed for fruit growers. The maximum working depth is limited to 55 cm. These machines can also be fitted with verticutting tools, so as to be able to aerate the soil down to a depth of 35 cm.

MECHANICAL TRENCHING WITH A BULLDOZER OR TRACTOR (TRENCH PLOUGH).   In trenching, the soil is loosened and turned with a special plough, down to a depth of 60 cm or even more. This operation brings dead soil from deeper soil layers into the upper layers, and as a result the humus-rich topsoil, with large numbers of aerobic organisms, is buried. This leads to a decrease in active soil life. Under no circumstances should trenching be carried out if the soil is healthy: a topsoil which has a high content of humus and living organisms must not be buried.

In private gardens (e.g. after compaction during house construction), trenching can also be carried out manually (20–60 m^2 per person per day).

Fig. 2.19. Good root growth in optimally prepared soils.

Organic preparation of the soil by green manuring or fallow systems

Deep loosening of compacted soils is advisable only if deep-rooting **green manure plants** are sown afterwards. The roots grow through the layers of soil that have been broken up and thus prevent recompaction of the soil.

ADVANTAGES OF GREEN MANURING

- Stabilization of the soil
- makes nutrient humus available
- development of soil organisms
- nitrogen enrichment (legumes)
- soil is loosened by the roots
- deeper-lying nutrients are made available (P)
- provides shade and thus optimizes soil tilth
- inhibition of weed growth
- protection from erosion
- reduction of soil exhaustion (inhibition of nematodes).

GREEN FALLOW.　If there are replanting problems (poor tree growth due to soil exhaustion), a green fallow for a few years is advisable. Complete destruction of the carriers of viral diseases and other microorganisms (e.g. actinomycetes as a cause of soil exhaustion) is not always possible with a green fallow (Rüdel, 1989).

The following effects are known:

- Cultivation of perennial plants does not inhibit nematodes but instead increases them at a depth of 60 cm.
- Most inhibitory substances are degraded by strong biological activation of the soil; the soil structure is improved and the soil is enriched with nitrogen and organic matter.
- There is therefore no need for initial nitrogen fertilization in newly planted orchards.
- Oil radish and mustard, but not cereals, are beneficial in a 1-year fallow.
- A 1-year green fallow is less advisable, as it does not control nematodes or give any significant improvement in soil structure.

A green fallow must be continued for several years if the maximum benefit is to be obtained from it.

GREEN FALLOW AND GREEN MANURE MIXTURES.　The following mixture has given good results in difficult, dense clay soils after loosening of the soil:

maize	20 kg/ha
horsebeans	20 kg/ha

vetch or lupin	20 kg/ha
peas	20 kg/ha
sunflowers	5 kg/ha
fodder oats	50 kg/ha
total	**135 kg/ha**

Sowing takes place between the **middle and end of April**. Under normal weather conditions, a lush plant cover develops by the end of July. This is then chopped up with a flail cutter and left to wilt for 2–3 days, after which it is ploughed in. The second plant cover, **15–20 kg/ha of oil radish** or **field mustard**, is then sown at the beginning of August.

The following **nematode-inhibiting mixture** should be used for the winter cover crop:

phacelia	4.0 kg/ha	
buckwheat	4.0 kg/ha	
berseem	5.0 kg/ha	Sowing before **mid-August** is
Persian clover	2.0 kg/ha	recommended.
Maxi white mustard	2.0 kg/ha	
Pegletta oil radish	3.0 kg/ha	
total	**20.0 kg/ha**	

A drought-tolerant mixture, based predominantly on legumes, which provides additional **nitrogen for the subsequent crop**, has also given good results as the winter cover crop:

flat peas	20.0 kg/ha	
seed vetch	20.0 kg/ha	
buckwheat	2.0 kg/ha	
berseem	4.0 kg/ha	Sowing should be completed by
mallow	1.0 kg/ha	**mid-August** at the latest.
phacelia	2.0 kg/ha	
mustard	1.0 kg/ha	
total	**50.0 kg/ha**	

Wild-flower strip mixtures for increasing biological diversity and encouraging important beneficials in orchards

Wild-flower strip mixtures can be used for the long-term establishment of wild-flower strips between the rows and at the edge of the orchard. This is a highly effective way of achieving plant and animal species diversity and encouraging important beneficials in orchards. Typical mixtures include wild-flowers such as borage, camomile, various types of clover, cornflower and yarrow.

Biocoenosis

This means the interaction of fruit plants with each other and the inter-action of fruit plants with the fauna and flora present. In organic fruit growing, the aim is to increase biological diversity, i.e. provide a biotope for a rich fauna and flora. An organically managed orchard presents an ideal combination of different ecological elements.

Herb-rich tramlines: undergrowth with a wide diversity of species and especially herbs, low-growing, drought-tolerant and providing good access for vehicles. Cutting should be alternated, to ensure that flow-ering grasses and herbs are always present in the orchard.

Rows with a wide diversity of weeds: the rows should not be kept clear throughout the year; even with young trees, some ground cover should be tolerated, at least in winter.

Wild-flower strips: selected wild-flowers are sown as perennial strips on the edge or within the orchard.

Extensive meadows in unused areas: unproductive, relatively undis-turbed areas (e.g. edge strips, small unused areas, edges of paths, slopes) should be converted to extensive meadows with a wide diversity of species.

Planted hedges: hedges provide a biotope and refuge for plants and ani-mals which have often already become rare, as well as for natural enemies of pests (e.g. birds, hedgehogs, beneficial insect species), pollinators (e.g. bees) and complex communities which enhance the ecological balance.

Individual trees (fruit trees and other tree species which are appropriate for the location): orchards consisting of standard trees in the vicinity of commercial orchards should be left or replanted. Untreated apple trees are a good reservoir for predatory mites and are inhabited by over 1000 species of insects, mites and spiders. Standard fruit trees are an important element of the landscape and rural image.

Artificial refuges: nesting boxes for birds; piles of stones and branches as places for insects, arachnids, small mammals and reptiles to spend the winter; perches for birds of prey.

Planting stock for an organic orchard

Under the EU guidelines for organic production, only planting stock obtained by the organic production method may be used in future (except for transitional periods). This also presupposes that nurseries must adapt accordingly and plan their production to conform with the guidelines for organic production; i.e. the rootstocks and young trees have to be grown in organic nurseries.

Propagation of planting stock in the organic nursery

Choice of site (location and soil)

Only relatively flat sites are suitable for nurseries. On sloping sites there is a risk of erosion because the soil has to be kept bare. In addition, locations where there is a risk of frost (winter frost and late frost) should not be used as nurseries, especially for the propagation of stone fruit and pears on quince rootstocks. Nurseries should preferably be set up on medium–heavy soils. Humus-rich, loamy sand soils or sandy loam soils are especially favourable. A soil index well above 50 is advisable for nurseries.

Soil type

- **Sandy soil**: can be tilled early – it is possible to till the soil soon after rain (no danger of compaction). Only suitable for a nursery if artificially irrigated.
- **Heavy soils**: not suitable, because they are too wet and cold.
- **Medium soils**: are the **most suitable** for nursery use. They have good aeration and a good storage capacity for water and nutrients.

The humus content should be between 1.5 and 4%, and the pH between 5.5 and 6.5. If the humus content is too low, it is advisable to enrich the soil with organic matter (see section on *Organic fertilizers for soil conditioning* in Chapter 4), and if the pH is too low the soil should be limed with dolomite lime or calcium carbonate (2000 kg/ha).

Preparation of the soil

An analysis of the soil is necessary before setting up an organic nursery. The level of basal dressing is decided on the basis of this. After the basal dressing and any other soil-improvement measures needed (deep loosening), it is advisable to sow a green manure. Nitrogen-fixing green manure plants (legumes) should preferably be used to optimize the nitrogen supply. A winter ground cover of legumes, which makes nitrogen available for the subsequent crop (young trees), should be grown after this green manure. Sowing should be completed by mid-August at the latest. It is advisable to apply organic nitrogen fertilizers before the soil is tilled in the spring. Oilseed meals (rapeseed or castor oilmeal), slurry, molasses, meat meals, etc., are especially suitable for this purpose.

Tilling the soil before planting out

- Plough (about 20 cm) after prior humus enrichment with green manure or the grower's own humus fertilizer (combined with basal dressing if necessary).
- Till the soil with a rotary cultivator just before planting out.

Planting out

Preparation of rootstocks or grafts

- Shorten the roots (10–14 days before planting out) with a hoe and dip in a copper solution (1%) as a precaution against crown gall.
- Cut back to a height of about 50 cm (measured from the root collar) before planting out the rootstock, except in the case of cherry rootstocks.
- Dipping in a clay poultice or putting in water produces better growth.
- **Spacing**: 80–120 cm × 30–40 cm, depending on planting stock. It is absolutely essential to keep to this spacing, in order to achieve a good crop of early shoots. In determining the spacing in the nursery, however, the tools available for weed control (maize chopper) should also be taken into account.

Technique for planting out

- Planting out **by hand**: spade
- **semi-mechanical** planting out: with drill markers or simple planting tools.

DRILL MARKERS. These should not be considered unless the number of rootstocks is at least 10,000–15,000. They can be attached to a tractor and used to make furrows in the soil with a selected row spacing. The rootstocks are then put into the furrows, trodden in and earthed up.

PLANTING TOOLS FOR ROOTSTOCKS AND GRAFTS. A one- to three-row planting tool is used; a share makes a furrow, the rootstocks are put into the furrows by a single person, and the planting tool then firms them in and earths them up.

Fertilizers (also see section on Important organic fertilizers for fruit growing: Chapter 4)

- **Continuous N fertilization**: oilseed meals (rapeseed or castor oilmeal), slurry, molasses, meat meals, Biosol, Biofert.
- **Foliage dressing sprays** (usually combined with organic insecticides and/or organic fungicides) – if required – should not be used until a few leaves have developed.

Watering

An adequate water supply is needed in the first few months, to ensure that the soil settles properly and the rootstocks show good root growth. Watering is absolutely essential on sandy soils.

Continuous control of diseases and pests (also see Chapter 5 on Plant protection)

It is of particular importance to watch out for any signs of scab and mildew, as well as aphids, red spider mite and rust mite. Young splice or bud grafts which are starting to sprout can be damaged by the grapevine weevil (*Rhynchites conicus*). In some areas of Europe it is absolutely essential to check the nursery for voles before winter dormancy (catch them with traps or fumigate the burrows).

Loosening the soil and controlling weeds

Since the use of herbicides is not appropriate in organic production, weeds are controlled with **rotary cultivators** or **cultivator attachments**. This is best done after planting out, because the soil is often severely compacted. Early weed control is especially important in the first few months.

Quality of planting stock

One of the most important factors for the success of an orchard is the quality of the planting stock. In addition to certain basic requirements, the planting stock must be of satisfactory internal and external quality. Nowadays it is unusual to plant trees without lateral shoots. With these trees, the fruit grower is acting as a nurseryman in the first year, as he has to grow the lateral shoots in his orchard. This inevitably reduces the initial yields.

In organic fruit production, in particular, it is essential to use only planting stock of the highest quality, since it is not possible to use chemical aids to improve tree development. The motto for organic fruit growers must therefore be: '**Only the best is good enough**': in other words, only properly prepared planting stock (premature or 1-year shoots) should be used.

This means that only good-quality bud grafts, 'knip' trees or normal 2-year-old trees should be considered as planting stock.

Table 2.11. Minimum requirements for the planting stock.

healthy	free of diseases (e.g. fruit tree canker, crown gall) and pests (e.g. San José scale, red spider mite), virus-free, free of mechanical damage and damage due to weather (e.g. hail, frost)
vigorous	adequate strength and length
uniform	same age, same rootstock, uniform quality
true to rootstock	uniform rootstock or rootstock type or clone
true to variety	all the trees correspond to the variety stated on the label
varietally pure	no mixing of varieties
virus-free	free of all known viruses, by thermotherapy

Fig. 2.20. Poor tree material delays the start of cropping.

Fig. 2.21. High-quality planting stock with a lot of premature shoots.

Criteria for high-quality planting stock

Rootstocks (1-year-old shoots from stoolbeds or transplants)

- At least 6 mm root collar diameter
- straight shoots
- good root system (at least six roots)
- at least 20 cm length from the root collar to the first branching.

Grading: 5–7 mm (measured at a height of 30 cm)
7–9 mm (for bud grafts)
9–12 mm (for splice grafts)

One-year-old bud grafts

- Adequate root system
- straight stem with good wood maturity
- stem diameter 10 cm, at least 13 mm above the graft union
- at least four premature shoots (>30 cm) at a height of at least 70–100 cm
- shallow angle
- minimum length 1.0 m (preferably 1.30 m)
- height of graft union at least 15 cm
- length of rootstock at least 30 cm.

Two-year-old trees with a crown consisting of feathers ('knip' trees)

- Cut height at least 50 cm (preferably 60 cm)
- stem diameter 10 cm, at least 13 mm above the graft union
- at least four feathers (>30 cm) at a height of at least 70–100 cm
- shallow angle
- minimum length 1.0 m (preferably 1.30 m)
- height of graft union at least 15 cm
- length of rootstock at least 30 cm.

Normal 2-year-old trees

- Cut height between 80 cm and 100 cm
- height of the ramification in the range at least 70–100 cm
- at least four 1-year-old shoots >30 cm (laterals competing with the leader do not count)
- shallow angle
- stem diameter 10 cm, at least 13 mm above the graft union
- minimum length 1.30 m
- straight stem with good wood maturity.

Planting systems in organic fruit production

There are not yet many specialized fruit growers using organic production methods; most are mixed enterprises, usually with small, extensive fruit-growing areas. The switch to low standard systems took place later than in integrated production. The planting density has also increased in recent years, but is still lower on average than in integrated production.

The greater planting distances between the rows and within the rows are intended to achieve better ventilation and exposure to light. In organic production with more extensive orchards, the yields are often lower and the manual labour requirements are higher, because of lower planting densities and the high tree forms.

Significant improvements in absolute and relative productivity (yield per unit of area and costs per kilogram of fruit) can be expected from modern planting systems which are, in principle, not incompatible with the aims of organic agriculture. The higher manual labour requirements in organic orchards (especially manual thinning) can be reduced by using small tree forms, for example. It is important to reduce wire support costs (as few posts as possible) and to use materials which do not cause harm to the environment (e.g. no tropical wood).

Orchards with more than 4000–5000 trees/ha have very high establishment costs and increase the production risk. In many cases they do not achieve economically viable yields and make it difficult to produce quality fruit. In increasing the planting density on dwarfing rootstocks, the ecological requirements should be taken into account as well as the economic criteria. In organic production, for example, the aim is to achieve ground cover within the row of trees for a limited time, which could start a trend towards more robust rootstocks.

With small-crown tree forms and higher planting densities, the aims in organic production are similar to those in integrated production:

- early commencement of cropping
- regular yields through avoidance of biennial bearing
- improvement in fruit quality through optimum light exposure
- reduction in manual labour requirements (pruning, thinning, harvesting)
- efficient mechanization (mulching, plant protection, soil management)
- easier protection of crops from frost and hail (anti-frost irrigation, hail nets)
- reduction in production costs
- rapid adaptation of the range to consumer demands and to advances in tree breeding (resistant varieties).

The following aspects should be borne in mind in choosing **planting distances**:

- The more dwarfing the rootstocks and the shorter the planting distances, the smaller is the useful tree volume per tree and the greater the sensitivity of the orchard to shortages of water and nutrients (Weibel, 1995). Fertigation and foliar fertilization for short-term correction of the nutrient supply are permitted only to a limited extent in organic production, and only certain approved organic fertilizers can be used.
- The management of the soil is important in choosing the planting distances. If the distances between the trees are too short (<1.20 m), mechanized work on the row in the first few years after planting is difficult, as is also the mulching of rows with ground cover with the feeler arm. Shorter planting distances in the row necessitate other, usually expensive, methods of organically compatible soil management, such as covering with organic materials, mulch film or modern special equipment.
- Particular attention should be paid to measures for preventive control of diseases and pests. This is especially important in organic production, so loose plantings are advantageous. A single-row system with appropriate planting distances is generally suitable for organic production.
- The direction of the rows and the planting systems should be planned so as to achieve the optimum insolation (north–south direction). Nets to protect against hail also reduce the amount of light reaching the trees and thus decrease the development of fruit colour. Smaller trees grown as spindles produce little shading, especially if the distance between the tramlines is double the selected tree height **(optimum tree height = distance between rows/2 + 1.0 m).**

Recommended planting distances for different fruit species

Apple

Slender spindle/M9, single row	3.0–3.5 m × 1.0–1.3 m
Slender spindle/M26, M7, M11, single row (vigorous combinations)	3.5–4.0 m × 1.3–1.5 m

Pear

Slender spindle/quince MC, single row	3.0–4.0 m × 1.3–2.3 m
Adams quince/quince BA29, OHF333	4.0–4.5 m × 2.3–2.8 m

Plums

Slender spindle (J. Fereley, INRA GF 655/2)	3.5–4.5 m × 1.5–2.5 m

3

Choice of rootstocks and cultivars in organic fruit production

Dessert apple production

Choice of rootstock

Whereas 15–20 years ago priority was given to robust, usually vigorous rootstocks (e.g. Hohlkrone A2 and seedling rootstock), for economic reasons the emphasis today is on dwarfing rootstocks which are always propagated vegetatively. Apart from the well-known dwarfing rootstocks (M9, M26), semi-vigorous to vigorous rootstocks (M7, M4, M11) may also be suitable for organic production, as in some growing seasons a ground cover is grown on the planting strip. From about the third year after planting, more vigorous rootstocks are less susceptible to shortages of water and nutrients and are therefore more competitive. Against this there is the disadvantage of often too vigorous shoot growth and a later commencement of cropping.

Dwarfing rootstocks

M9 (GELBER METZER PARADIES). M9 is the rootstock which is most widely used in commercial fruit production. The advantages of this rootstock are its early fertility and its dwarfing habit. It is thus the ideal rootstock for small-crowned tree forms. M9 is very sensitive to lack of oxygen, and on poor sites (with waterlogging) the rootstock tends to early exhaustion. M9 needs very good soil. There are a large number of M9 clone rootstocks, which vary widely in vigour.

M26. This rootstock is less demanding than M9 in terms of soil requirements and has rather better anchorage. M26 virus-free is about 50% more vigorous than M9, and this is why the commencement of cropping is later.

Semi-vigorous rootstocks

M4.   Suitable for bigger crowns, not sufficiently well anchored and does not tolerate either dry soils or soils that have a variable moisture content or are subject to waterlogging.

M7.   Also suitable for heavy, wet soils; replant rootstock for rotation areas with soil exhaustion; suitable both for spindle forms and for other crown forms.

MM 106.   Early commencement of cropping; only suitable for dry locations, as it is susceptible to collar rot.

Vigorous rootstocks

M11.   Broader crown, susceptible to crown gall, positive effect on fruit quality, no special soil requirements.

M25.   Vigour 80% of the seedling rootstock, early commencement of cropping, positive effect on fruit quality and commencement of cropping; the rootstock is not sufficiently well anchored and is suitable for more extensive use (domestic gardens, cider apple production).

New rootstocks for commercial fruit production

Until recently M9 was almost the only rootstock used for apples, but in the last few years, just as for cultivars, the range of rootstocks available has greatly increased. The reason for the development of new rootstocks is that the M rootstocks, especially M9, not only have advantages but also have major disadvantages, such as poor anchorage, relatively low frost hardiness, susceptibility to drought, susceptibility to fireblight, tendency to attack by woolly aphids and rodents, and the occurrence of aerial roots and suckers. Under certain conditions (virgin soil, vigorous cultivars) M9 has often proved to be too vigorous, but sometimes also too dwarfing. The aims of current rootstock breeding are therefore to combine as many positive properties as possible in one rootstock (control of shoot growth, rapid commencement of cropping, reduction in biennial bearing, improvement in fruit quality, and resistance to winter cold, diseases and pests). If there is a danger of fireblight, not only the cultivar but also the rootstock should be resistant to this disease.

M9 CLONE ROOTSTOCKS.   Much breeding work at present is concentrated on the M9 rootstock. These days there is a range of M9 selections exhibiting different behaviour in the nursery (root system, growth, stoolbed performance, etc.).

Table 3.1. The most important M9 clone rootstocks.

Rootstock	Origin	Growth	Remarks
T337	The Netherlands, NAKB[a]	M9	standard rootstock in Europe
Fleuren 56	The Netherlands, Fleuren Nursery	< M9 T337	very high productivity, enhancement of colour development, for virgin soils
M9 EMLA	UK	> M9 T337	first virus-free M9 rootstock
Pajam 1 (Lancep)	France, CTIFL[b]	like M9 T337	high productivity, virgin soils
Pajam 2 (Cepiland)	France, CTIFL	> M9 T337	more vigorous M9, for replant
Nicolai clone (e.g. Nic 8, 13, 19, 29)	Belgium	> M9 T337	very productive, early commencement to cropping, vigorous, some runners
Burgmer clone (B719, 751, 984, 756)	Germany, Burgmer Nursery	> or < M9 T337	good productivity, virus-free

[a] NAKB: General Netherlands Inspection Service of Woody Nursery Stock.
[b] CTIFL: Centre Technique Interprofessional des Fruits et Légumes.

There are large differences in performance between the various selected M9 clone rootstocks. Their origin is a critical factor. M9 clones generally perform better and are to be preferred. Freedom from viruses is essential for the vitality of a rootstock. **Virus-free rootstocks** are superior in all respects:

- they show healthier growth and are more suitable for replanting
- they give higher yields
- cropping starts earlier
- the colour development of the fruit begins earlier and is more pronounced
- runner formation is limited.

Table 3.2. Example Elstar-Elshof; planted winter 1992/93.

Rootstock	kg per tree 1994–96	Fruit weight (g)	Stem circumference (cm) March 1997
T337-15cm*	43.0	180	15.1
Fl 56	34.1	177	13.3
Nicolai 29	40.0	181	15.2
Jork 9	28.2	162	10.8

* Graft height.

Choice of cultivars

Choice of the cultivar is of particular importance in organic fruit production; for the grower it is primarily a question of economics, because the useful life of the orchard extends for a number of years. The choice of cultivar is greatly influenced by the site, by business and sales considerations, and by the personal inclinations of the grower. To achieve economic success there must be an optimum match between the site requirements of the cultivar and the site conditions (**choice of the cultivar that is appropriate for the site,** i.e. 'the right cultivar in the right location'). The use of chemical aids needs to be reduced by taking advantage of specific cultivar properties, such as resistance to diseases, pests, physiological disturbances and weather effects (**choice of the cultivar that is appropriate in organic terms**). The susceptibility of cultivars to disease and pests is of vital importance in organic fruit production, with limited possibilities for the use of chemical aids. There is as yet very little experience of the use of modern market cultivars in organic fruit production; there are also regional differences in the performance of cultivars in terms of yield, quality, and susceptibility to diseases and pests. To optimize the choice of cultivars it is very important, therefore, to obtain regional information on specific cultivar properties through intensive testing, i.e. before new cultivars are planted out, they should be tested for their commercial suitability in regional testing stations.

The choice of cultivars and the range of cultivars in the commercial orchard are primarily determined by the type of sale:

- direct sale (sale from the orchard and local markets)
- sale via wholesalers to retail chains.

In choosing cultivars, care should be taken to ensure that cultivar properties are in accord with the site and the business structure. In addition, it is important not to grow too many cultivars, making the workload at harvest time unmanageable. At the same time, however, **the market risk should be spread over more than one cultivar**. It is also important for every orchard to have a spectrum of cultivars appropriate for its site (early, autumn and storage cultivars). Fruit growers also need to bear in mind the likelihood that they will no longer be able to market their organic products solely by direct sale but will increasingly need to win custom from the retail chains. This means that the requirements for cultivars in organic production will be similar to those in integrated production.

Description of the most important market cultivars for organic production

In the following descriptions, **V** indicates a cultivar with varietal protection, ® indicates a cultivar with trade mark protection.

Table 3.3. Assessment of early cultivars (summer cultivars) for organic growing, ripening July–August.

Classification	Cultivar	Market prospects	Recommended?
main cultivars	Summerred	good–poor	no, susceptible to scab
	Delbarestivale	good	yes, red mutants (Celeste, Ambassy)
subsidiary cultivars	Discovery	local market	
	Vista Bella	poor	no, reduce
	Jerseymac	poor	no, greatly reduce
new cultivars	Astramel		do not plant
	Sommerregent	local market	yes, to replace James Grieve
	Early Gold	local market	observe
	Arkcham		organic cultivar
	Sunrise		observe
	Piros		organic cultivar
	Retina		resistant cultivar
declining or discontinued cultivars	James Grieve	zero	remove
	Gravensteiner	local market	
	Mantet	zero	
	others		remove

Table 3.4. Assessment of autumn and winter cultivars for organic growing, ripening September.

Classification	Cultivar	Market prospects	Recommended?
main cultivars	Elstar	good	yes, expand
	Gala	very good	no, susceptible
	Kronprinz Rudolf	good	yes, local cultivar
	McIntosh		no, (greatly) reduce
subsidiary cultivars	Arlet	local market?	no, scab
	Roter Boskoop	good	yes, late locations
	Rubinette	good	no, scab
	Greensleeves	poor	organic cultivar
	Alkmene	local market	robust cultivar
	Fiesta	poor	yes, low susceptibility
new cultivars	Ecolette		resistant to scab
	Santana		resistant to scab, worth a trial
	Rubinola		resistant to scab, worth a trial
	Rosana		resistant to scab
	Rajka		resistant to scab
	Resi		resistant to scab, worth a trial
	Reanda	processing cultivar	resistant to scab
	Relinda	cider apple cultivar	resistant to scab
	Reglindis	processing cultivar	resistant to scab
	Remo	cider apple cultivar	
declining or discontinued cultivars	Lobo	poor	no, reduce
	Empire	poor	remove
	others		remove

Table 3.5. Assessment of autumn and winter cultivars, ripening October.

Classification	Cultivar	Market prospects	Recommended?
main cultivars	Golden Delicious	moderate (quality)	no, susceptible
	Idared	good	yes, warm locations
	Jonagold	good	dependent on conditions, red mutants
	Gloster		no, reduce
subsidiary cultivars	Granny Smith	poor	no, do not expand
	Florina	local market	organic cultivar
	Delbard Jubile	local market	
	Braeburn	good	no, susceptible
	Fuji	local market	no, susceptible
new cultivars	Meran		organic cultivar
	Elise	good	worth a trial
	Pinova	good	worth a trial
	Topaz	good	organic cultivar, worth a trial
	Otava		worth a trial
	Goldstar		worth a trial
	Delorina		worth a trial
	Enterprise		worth a trial
	Goldrush		warm locations, late cultivar
declining or discontinued cultivars	Red Delicious	poor	remove
	Jonathan	local market	reduce
	Mutsu	poor	remove

DELBARESTIVALE® (DELCORF)

Origin: Delbard Nursery, France: Stark Jongrimes × Golden Delicious cross

Ripening: mid–end August, spread out over a long period, more than one picking needed, tendency to early fruit drop, colour development often inadequate

Fruit: medium size, cylindrical shape with a deep, wide calycinal cavity, bright red or red stripes on a yellowish green background

Taste: very juicy, crisp, good aroma, sweet

Yield: early cropping, moderately high yield, because of risk of biennial bearing

Growth: semi-vigorous to vigorous

Susceptibility: some susceptibility to scab and mildew, early fruit drop

Keeping quality: satisfactory, until the end of October in cold storage at 3°C

Red mutants: Ambassy® (Dalili), Davodeau Ligonniere Nursery (France)
                 Celeste (improved standard), Herr Nursery (Germany)

SOMMERREGENT

Origin: Germany; Anton Fischer × James Grieve cross

Ripening: mid–end August (5–6 days before James Grieve), spread out over a long period, more than one picking needed

Fruit: medium size, agreeable bright red or red stripes on a yellowish green background

Taste: very juicy, crisp, with no particular aroma
Yield: early cropping, high yield, no biennial bearing
Growth: extremely weak, use more vigorous rootstocks
Susceptibility: low, extremely robust cultivar
Keeping quality: satisfactory, better than James Grieve in cold storage

PIROS (V)
Origin: Dresden Pillnitz, Germany; Helios × Apollo cross
Ripening: end of July
Fruit: medium to large, agreeable red stripes on a yellowish green background
Taste: good to very good, moderately firm, juicy, tart with good aroma
Yield: early cropping, medium-high yield, tendency to biennial bearing
Growth: weak to semi-vigorous, little ramification
Susceptibility: low, extremely robust cultivar, moderately susceptible to fireblight
Keeping quality: 3–4 weeks in cold storage

ELSTAR (V)
Origin: CPRO-DLO Wageningen, The Netherlands; Golden Delicious × Ingrid Marie cross
Ripening: early to mid-September, colour development often inadequate at the optimum picking time
Fruit: medium size with shiny red striped colour
Taste: very juicy, crisp, pungent fruity aroma with a lot of sugar and refreshing tartness, also highly suitable for processing
Yield: early cropping, moderately high yield, because highly susceptible to biennial bearing
Growth: vigorous to very vigorous, strong ramification
Susceptibility: some susceptibility to scab and moderate susceptibility to mildew, biennial bearing
Keeping quality: good, until April in controlled-atmosphere storage, tends to become soft when taken out of storage
Red mutants: Elstar (V)
　　　　　　　Red Elstar (V)
　　　　　　　Red Elswout®

FIESTA (V)
Origin: East Malling, UK; Cox Orange × Idared cross
Ripening: early to mid-September, tendency to early fruit drop
Fruit: medium size with bright red or orange-red striped colour, flattened sphere shape
Taste: satisfactory to good, firm, crisp and juicy, aromatic
Yield: early cropping, high yield, some tendency to biennial bearing
Growth: semi-vigorous, good ramification
Susceptibility: fruit tree canker and fireblight
Keeping quality: good, until the end of May in controlled-atmosphere storage, good shelf life

BOSKOOP
Origin: Boskoop, The Netherlands; pedigree unknown

Ripening: end of September
Fruit: large with dull, usually russeted, skin
Taste: fruit moderately firm, very tart, good for cooking and baking
Yield: moderately early cropping, moderately high yield, biennial bearing
Growth: vigorous to very vigorous
Susceptibility: frost damage to flowers, bitter pit
Keeping quality: good, until mid-May in controlled-atmosphere storage, do not store <3°C
Red mutants: Roter Boskoop Schmitz Hübsch
                Bielaar®

KRONPRINZ RUDOLF
Origin: random seedling, Styria, Austria, around 1860
Ripening: end of September
Fruit: small to medium size, flattened sphere shape, colour one-third to two-thirds bright carmine red
Taste: flesh white, moderately firm, sweet and sharp, juicy
Yield: moderately early cropping, medium to high yield, biennial bearing, good thinning needed
Growth: vigorous to very vigorous
Susceptibility: susceptible to scab, sensitive to pressure
Keeping quality: good, until mid-March in controlled-atmosphere storage, some superficial scald

JONAGOLD
Origin: Geneva Experiment Station, New York State, USA; Golden Del. × Jonathan cross
Ripening: mid- to end September, spread out over a long period
Fruit: large to very large, shiny red colour, striped to pale, depending on type
Taste: flesh moderately firm, sometimes friable, juicy, very aromatic and well-balanced in flavour
Yield: early cropping, sometimes slight tendency to biennial bearing
Growth: initially very vigorous, later semi-vigorous
Susceptibility: bitter pit, some susceptibility to scab
Keeping quality: good, until mid-June in controlled-atmosphere storage
Red mutants: Jonagored (V)
                Novajo (V)

PINOVA (V)
Origin: Dresden Pillnitz, Germany; Clivia × Golden Delicious cross
Ripening: early October (like Golden Delicious), more than one picking needed, delayed colour development
Fruit: medium size, agreeably striped bright orange-red on yellowish green background
Taste: very firm, juicy, crisp with satisfying aroma
Yield: early cropping, regular and very high yield, no biennial bearing
Growth: weak, good ramification
Susceptibility: low to scab, robust cultivar, apart from fireblight (second flowering) and mildew

Keeping quality: very good, 6–7 months in controlled-atmosphere storage, long shelf life

ELISE (V)
Origin: CPRO-DLO Wageningen, The Netherlands; Septer × Cox Orange cross
Ripening: early October, do not pick too early
Fruit: medium size to large, elongated, three-quarters of the fruit dark red over a large area
Taste: firm, juicy, aromatic
Yield: early cropping, regular high yield
Growth: semi-vigorous, good ramification
Susceptibility: fruit tree canker, lenticel blotch pit
Keeping quality: good, 6–7 months in controlled-atmosphere storage

DELBARD JUBILE®, DELLGOLLUNE (V)
Origin: Delbard Nursery, France; Golden Delicious × Lundbytrop cross
Ripening: end September–early October, spread out over a period, more than one picking needed
Fruit: large to very large, mottled bright red
Taste: flesh moderately firm to soft, juicy, sweet, good aroma
Yield: early cropping, medium yield because of risk of biennial bearing
Growth: vigorous, moderate ramification
Susceptibility: little scab and mildew, sensitive to sulphur
Keeping quality: good, until the end of May in controlled-atmosphere storage

IDARED
Origin: Idaho Experiment Station, USA; Jonathan × Wagener cross
Ripening: beginning of October, do not pick too early
Fruit: medium size to large, colour bright red
Taste: firm, juicy, delicate mild flavour, good for cooking and baking
Yield: early cropping, regular high yield
Growth: semi-vigorous to weak growth
Susceptibility: mildew, highly susceptible to fireblight
Keeping quality: good, until the end of May in controlled-atmosphere storage, do not store <3°C

MERAN (V)
Origin: South Tyrol, random seedling
Ripening: start of October; more than one picking needed, delayed colour development
Fruit: medium size to small, with shiny mottled colour
Taste: very firm, juicy, delicately crisp with satisfying aroma
Yield: early cropping, regular very high yield, no biennial bearing
Growth: semi-vigorous, good ramification
Susceptibility: low susceptibility, robust cultivar under local production conditions
Keeping quality: very good, until the end of June in controlled-atmosphere storage

Resistant apple cultivars

Various wild species of the *Malus* genus, as well as resistant cultivars, are used in breeding to confer resistance.

Table 3.6. Sources of resistance: wild and ornamental apple species.

Mildew	*Malus zumi*
	Malus robusta
Scab	*Malus floribunda* (V_f)
	M. pumila (V_r)
	M. micromalus (V_m)
	Antonovka (V_a)
Fireblight	*M. floribunda*
	M. robusta
	Old Home (pear)
Aphid	*M. robusta*
Woolly aphid	Northern Spy
Rodents	*M. prunifolia*
	M. sieboldi

So-called V_a **resistance** (Antonovka), V_r **resistance** (*Malus pumila*) and especially V_f **resistance** (*Malus floribunda*) are bred into the scab-resistant apple cultivars. Most cultivars with V_f resistance are descended from a crossing made in the USA at the beginning of the 20th century between *Malus floribunda* clone 821 and Rome Beauty. Previously it was assumed that V_f resistance was monogenic (determined by a single gene), whereas the other resistance mechanisms were polygenic (determined by a large number of genes). Recent findings, however, suggest that V_f resistance is not monogenic either. Scab has already been found at various locations on cultivars which have V_f resistance. It is thought that part of the basic genetic material which is responsible for the high resistance of *Malus floribunda* was replaced by genetic material from more susceptible cultivars during the breeding programme. In this way the resistance genes were weakened. The choice of a susceptible parent cultivar in the breeding of scab-resistant cultivars increases the number of non-resistant descendants. For this reason only robust cultivars should be used for cross-breeding. A number of studies have also shown that as a result of strong selection processes, the virulent scab strains become specialized on particular cultivars. Extensive cultivation of less scab-resistant cultivars with only one resistance mechanism therefore increases the risk of development of scab strains which are able to circumvent this resistance. Kellerhals and Gessler therefore advocate **mixed planting of cultivars** with different resistance mechanisms. Another possible approach is mixed planting of cultivars with similar ripening times that have low susceptibility or are resistant to scab.

With increasing cultivation of scab-resistant cultivars, however, it is important to bear in mind the risk of adaptation of the apple scab fungus (*Venturia inaequalis*). Even when using organic production methods, the cultivation of scab-resistant cultivars does not mean that fungicide treatment can be completely abandoned. In most regions there is a great increase in sooty blotch if treatment against apple scab is stopped completely.

INRA Angers breeding programme (France)

FLORINA (V). Origin **France**, trade name Querina®, cross-bred from *Malus floribunda* 821, Morgenduft, Starking Delicious, Golden Delicious and Jonathan; V_f-resistant, low susceptibility to mildew; moderate to good colour, dark red, pruinosity (wax coating) typical of the cultivar; moderate to good flavour quality, sweet and lightly perfumed; good yield and fruit size, susceptible to biennial bearing; many fruits with core flush.

DELORINA (V). Origin **France**, trade name Harmonie®, Blushing Golden × Florina cross; V_f-resistant, low susceptibility to mildew; fairly good colour development, lightly striped; firm flesh and good keeping quality, rather sensitive to cold (soft scald); heavy-bearing and good fruit size; moderate shoot growth, simple training.

CPRO-DLO breeding programme (The Netherlands)

ECOLETTE®. Origin **The Netherlands**, Elstar × Prima cross; V_f-resistant and low susceptibility to mildew; moderate to good, lightly striped bright red colour; good flavour quality, rather tart, firm flesh and good keeping quality; moderate yield and fruit size, fruit on 1-year-old wood in particular tends to be small; vigorous shoot growth with pronounced apical dominance.

SANTANA®. Origin **The Netherlands**, Elstar × Priscilla cross (CPRO 78038-9); good-sized fruit with good colour, sweet, sharp flavour; yields high and regular, vigorous shoot growth with tendency to defoliation, some susceptibility to mildew.

Strizovice u. Holovousy breeding programme (Czech Republic)

OTAVA (V). Origin **Czech Republic**, Shampion × Jolana cross; V_f-resistant; yellowish green background colour with a light-pink blush on the cheek, good flavour quality, rather tart; high productivity with tendency to small fruit size, little shoot growth; susceptible to magnesium deficiency.

RUBINOLA (V). Origin **Czech Republic**, Prima × Rubin cross; V_f-resistant; striped red colour, good flavour quality; low yield and small fruit size; vigorous shoot growth with strong tendency to defoliation, difficult to train.

ROSANA (V). Origin **Czech Republic**, Jolana × Lord Lambourne cross; V_f-resistant; lightly striped, reddish brown colour, poor eating quality, rather acid; good yield and fruit size.

TOPAZ (V). Origin **Czech Republic**, Rubin × Vanda cross, 1984; orange-red, medium-sized, firm-fleshed and aromatic cultivar with good keeping quality; ripens about 1 week before Golden Delicious; early-bearing, with good and regular yields; V_f scab resistance, little susceptibility to mildew.

RAJKA (V). Origin **Czech Republic**, Shampion × Katka (Jolana × Rubin) cross, 1983; bright red, medium-sized, firm-fleshed cultivar with sweet, sharp flavour and good keeping quality; ripens about 10 days before Golden Delicious; early-bearing, with good and regular yields; V_f scab resistance, little susceptibility to mildew.

GOLDSTAR (V). Origin **Czech Republic**, Rubin × Vanda cross; ripening in October; greenish yellow, medium-sized, firm-fleshed cultivar with sweet, sharp flavour and good keeping quality, with pink tinge on the side exposed to the sun; early-bearing, with good and regular yields; V_f scab resistance, tolerant to mildew.

Coop selections (USA)

Of the many Coop selections, **Goldrush** (Coop 38), **Pristine** (Coop 32) and **Enterprise** (Coop 30) appear to be the most interesting at the present time. Goldrush ripens very late. Pristine is a summer cultivar and Enterprise a storage cultivar.

ENTERPRISE (V). Origin **USA**, Selection No. Coop 32; multiple cross of *Malus floribunda* 821, Golden Delicious and McIntosh; V_f-resistant, some susceptibility to canker; dark-red colour and moderate flavour quality, high fertility and large fruit size; moderate shoot growth, sensitive to cold in storage.

GOLDRUSH (V). Origin **USA**, Selection No. Coop 38; multiple cross of *Malus floribunda* 821 and Golden Delicious; V_f-resistant, some susceptibility to mildew, highly susceptible to sooty blotch; yellowish green background colour with brownish red cheek (similar to Golden Delicious), tendency to russeting on fruit; good flavour quality with good fruit maturity, late ripening; high yield and fairly good fruit size, susceptible to biennial bearing; moderate shoot growth and tree simple to train, late leaf fall.

PRISTINE (V). Origin **USA**, Selection No. Coop 32; multiple cross of *Malus floribunda* 821 and other cultivars; V_f-resistant, low susceptibility to mildew; yellow background colour with orange cheek, moderate eating quality, tart; heavy-bearing, small fruit size; vigorous cultivar.

Selections from Ahrensburg (Germany)

At the start of the 1980s, crossing with TSR 15T3 and Elstar and with Prima × Klon 40 was carried out in Ahrensburg (Germany). Three selections from these have since been given a varietal name: **Ahra**, **Ahrista** and **Gerlinde**.

AHRISTA (V).   Origin **Germany**, Selection No. BFA 80/2-62, TSR 15T3 × Elstar cross; V_f-resistant, low susceptibility to mildew; strong, moderately striped red colour, good flavour quality, not very firm-fleshed; high yield and good fruit size, weak to moderate shoot growth.

GERLINDE (V).   Origin **Germany**, Selection No. BFA 80/4-34, Elstar × TSR 15T3 cross; V_f-resistant, low susceptibility to mildew; dark-red, lightly striped colour, good flavour quality; heavy-bearing with good fruit size, unfortunately rather sensitive to cold (soft scald); vigorous shoot growth, pendulous branches; only suitable for domestic gardens.

Dresden-Pillnitz breeding programme (Re-cultivars®) (Germany)

RETINA (V).   Apollo × BX 44/2 cross; ripening at beginning of September, medium yield, early-bearing; fruit medium size, elongated, light- to dark-red colour (mottled to striped); flavour quality medium to good, medium-firm, juicy, sweet and sharp; growth medium to vigorous; V_f-scab-resistant, little mildew; storage life 2 months in cold storage.

REANDA (V).   Derived from Clivia; ripening early to mid-September, yield medium to high and regular, early-bearing; fruit medium size, spherically elongated, three-quarter lightly mottled red colour; flavour quality moderate, firm-fleshed and very juicy; growth weak with tendency to defoliation; V_f scab resistant, little mildew; storage life in cold storage until the end of January; good processing cultivar.

REMO (V).   James Grieve × BX 44/14 cross; ripening mid-September, yield very high and regular, early-bearing; fruit medium size, oblong, wine-red colour (mottled to striped); fruit very juicy, firm-fleshed, strongly acid, flavour moderate; growth weak, medium ramification; V_f scab resistant, little mildew, resistant to fireblight; storage life 2 months in cold storage; promising as cider apple variety.

REGLINDIS (V).   James Grieve × BX 44/18 cross; ripening mid-September, yield medium to high, early-bearing; fruit medium size, spherical, bright red colour (lightly mottled); fruit juicy, delicately firm-fleshed, sweet, sharp taste, flavour medium; growth semi-vigorous with good ramification; V_f scab resistant, little mildew, little susceptibility to spider mites; storage life 3 months in cold storage; promising as processing cultivar.

RELINDA (V). Undine × BX 44/14 cross; ripening middle to end of September, yield high and regular, early-bearing; fruit medium size, spherical, bright-red colour; flesh juicy, medium-fine texture, firm, sharp taste, flavour moderate to medium, little aroma; growth semi-vigorous with good ramification, thin shoots; V_f scab resistant, little mildew, little susceptibility to spider mites; storage life in cold storage until February; promising as cider apple variety.

RESI (V). Derived from Clivia; ripening end of September, yield medium to high, early-bearing; fruit small to medium size, three-quarter red colour (mottled to striped), flavour quality medium; growth weak to semi-vigorous; V_f scab resistant, little mildew; storage life in cold storage until March.

REGINE (V). Cox Orange cross open pollinated × scab-resistant breeding line; ripening at start of October, weak to semi-vigorous growth, good ramification, resistant to scab, early-bearing; fruit medium size, two-thirds red coloured, prominent lenticels.

Breeding programme of the Fruit Growing Research Station at Vigalzone di Pergine, Trentino (Italy)

GOLDEN LASA (V). Ed Gould Golden × P.R.I. cross 1956-6; ripening end of September; scab-resistant, low susceptibility to apple mildew; vigorous shoot growth, early-bearing with medium productivity; fruit large, yellowish green background colour (Goldenersatz); flavour sweet but sharp, moderately firm flesh.

GOLDEN MIRA (V). Coop 17 × Perleberg cross; ripening at beginning of October, scab- and mildew-resistant; vigorous shoot growth, early-bearing with medium productivity; fruit medium-sized to large, yellowish green background colour (Goldenersatz) with orange-red tinge on the side exposed to the sun; flavour very sweet, juicy, moderately firm flesh, scented.

GOLDEN ORANGE (V). Ed Gould Golden × P.R.I. cross 1956-6; ripening at start of October, scab-resistant; fruit medium-sized to large, yellowish green background colour (Goldenersatz) with orange-red tinge on the side exposed to the sun; flavour sweet but sharp, moderately firm flesh, poor keeping quality.

Table 3.7. The most important new apple rootstocks.

Rootstock	Origin	Cross	Growth	Advantages	Disadvantages
P22	Poland	M9 × Antonovka	< M9	Very frost-hardy, high production capacity, good fruit quality (colour development)	Root at severe risk of drying out (tree failures), juvenile and adult forms
P16	Poland	M9 × Antonovka	< M9	Very high productivity, enhancement of fruit size, virus-free	Root suckers in the juvenile stage
J9 (Jork 9)	Germany	Seedling from M9	like M9	More favourable than M9 for moister sites, high productivity, shallow graft angle, better anchorage than M9, good fruit quality, good ramification and crown formation	Highly susceptible to fireblight
Supporter rootstocks (1, 2 and 3)	Germany	M9 × M. baccata or M9 × M. micromalus	Between M9 and M27	Easy to propagate and early-bearing, high productivity	
B9 (Budagovsky)	Russia	(M8 × Red Standard)	> M9	Frost-hardy, better colour development, resistant to collar rot	Less productive than M9, susceptible to fireblight and woolly aphid, Smaller fruit Rapid spring development
MARK	USA	M types × A2 × Robusta	> M9	Very productive, early-bearing, good adaptation, positive effect on fruit colour, good affinity, no root suckers	Susceptible to A. tumefaciens Very expensive
CG Series	USA	Seedlings from M. prunifolia	> or < M9	Resistance to fireblight, collar rot and winter cold	Smaller fruit
Ottawa 3 (O3)	Canada	Rubin × M9	> M9	No aerial roots	Difficult to propagate
Bemali	Sweden	M4 × Codlin	> M9		

Table 3.8. Assessment of value of fruit cultivars.

1. Value in use	**external quality** size shape colour skin characteristics (blemish-free) free of diseases and pests	**internal quality** nutritional value organoleptic value (taste, aroma, texture) absence of harmful substances (pesticides, etc.) health value (vitamins, etc.) processing value (fruit products)
2. Market value	**marketability** attractiveness creates desire distinctiveness	**suitability for storage** keeping quality sensitivity to temperature and CO_2 persistence of acid and firmness **ease of handling and transport** skin blemishes leading to lower quality grades (impact bruises) firm flesh long shelf-life

3. Cultivation value

cropping	early-bearing, abundant and regular yield (little biennial bearing), few pickings
susceptibility	low susceptibility to diseases (especially scab and mildew), pests and physiological disorders (bitter pit, brown flesh, etc.), little tendency to russeting, no fruit drop before the harvest
quality	high proportion of fruit in commercial class I (only 1–2 fruit/inflorescence) – so there is little expenditure on fruit thinning
training	crowns easy to train; crowns as small as possible with abundant short fruiting spurs, no tendency to defoliation and ageing, semi-vigorous to weak growth

Value in cultivation = Market value / Production costs.

Dessert pear production

In many fruit-growing areas the production of pears has been given very low priority. In recent years, however, the demand for pears has risen markedly, partly because prices were good and partly because production of apples alone was regarded as too risky. In addition – as in apple growing – there have been radical changes in cultural techniques.

Choice of site

For optimum fruit quality, almost all pear varieties need a warm site, rather similar to that required for wine grapes. If these conditions are not present, the fruit very often does not have the desired melting texture and has little flavour; in extreme cases the fruit may taste 'carroty'. This applies particularly to late-ripening cultivars. Pear trees are somewhat more sensitive to winter frost than apple trees, and are at severe risk from late frosts because flowering is about 10 days earlier.

Soil requirements are similar to those for apples. Pears, especially on a quince rootstock, suffer from chlorosis when the pH is high.

All varieties of pears, even storage varieties, can only be kept for a few weeks without cold storage. They become mealy or their flesh turns brown after a short period of optimum ripeness. This period is particularly short in summer and autumn cultivars. This is why pears are often preserved as tinned fruit, in addition to being eaten as fresh fruit. A few special varieties are used for juice production and for manufacture of fruit brandies. Although they have a lower vitamin content than apples, fully mature, well developed pears are delicious.

Choice of rootstock

In the cultivation of pears, as in the cultivation of apples, the choice of rootstock depends on the growth characteristics of the cultivars to be used for grafting. On good virgin sites Quince A and Quince Adams can be virtually discounted if the aim is to achieve maximum vigour, although combination with dwarfing cultivars such as Conference, Gute Luise or Alexander Lukas may be feasible if the graft union is rather higher – about 20 cm. The same applies to the Quince BA29 rootstock, which is to be preferred on alkaline soils. Table 3.9 shows the most important rootstocks.

On suitable sites, Quince C is in fact the only rootstock which meets the requirement of high planting densities combined with satisfactory conditions of exposure to light. Unfortunately, because of its sensitivity to frost, this rootstock is often not sufficiently available and is risky in many sites. This applies particularly to growing in the nursery, but also to the first 2 years after planting. As a matter of principle, this rootstock

Table 3.9. Pear rootstocks.

Rootstock	Resistance		Cold	Compat-ibility	Vigour	Runner production	Lime tolerance	Output per tree
	Fireblight	Pear decline						
Seedling	0	3	good	5	5	high	good	3
Provence Quince	0	5	moderate	3	3	medium	poor	4
BA29	0	5	moderate	3	3	medium	good	4
Quince A	0	5	moderate	2–3	2–3	medium	poor	4
OHF Seedling	5	5?	good	5	4	rare	good	3
Quince Sydow	0	5	moderate	3	2–3	medium	poor	4
Quince Adams	0	5	low	3	2	high	poor	4
Quince C	0	5	low	3	1.5–2	little	poor	4–5
Pyrodwarf	2	?	moderate	5	2.5	little	good	4

1 = unsatisfactory; 3 = satisfactory; 5 = very good.

must be planted as deep as possible. As all quince rootstocks are sensitive to frost, the level of the graft union should not be too high. As a result, even with the slightly more frost-hardy quinces, such as Quince A, Quince Adams or BA29, it is not possible to overcome the problem of excessive vigour solely by high grafting. Recently, increased compatibility problems have been encountered with the combination of Alexander Lukas and OHF hybrids. The reasons for this are not yet clear. Because of this, caution should be exercised when planting OHF rootstocks.

Double-working

As is well known, some pear cultivars show problems of incompatibility with quince rootstocks. This is connected with the fact that although the pear and the quince belong to the same family (*Rosaceae*), they are not in the same genus. This incompatibility is particularly marked with virus-infected cultivars and for example with Williams, Clapp's or Charneux. Double-working is essential in these cases. Moreover, this can curb vigour and thus improve the reliability of cropping. In a trial with the cultivar Doyenne du Comice it was found up to the fifth year that double-working with Conference as the interstock, for example, gave much better results than the direct combination of Quince A and Doyenne du Comice. This fact certainly merits greater consideration in future. The cultivars Beurré Hardy and Doyenne du Comice are used as interstocks at present.

Planting stock

At present we consider that the optimum planting stock is a well-developed, virus-free, 2-year-old tree on a Quince C rootstock. With some cultivars, e.g. Williams' Bon Chrétien, an interstock is needed. On poorer sites, or when replanting, it may be advisable to use Quince A or Quince Adams. In most cases Quince C should be given preference, however. These trees produce faster and grow less vigorously as a result, so the tree structure is simpler.

The 2-year-old graft should have 4–6 equal shoots of ±40 cm length in ±75 cm. These shoots are then used as the framework branches. If Quince A is used as the rootstock, the graft union should not be more than 20 cm above the ground. In the case of Quince C, 2 cm is sufficient, so as to provide better protection against winter frost.

Methods of training

An important guiding principle with all tree forms and all methods of training is that the slower and more controlled the upward growth of the centre of the tree, the less is the danger of overcrowding.

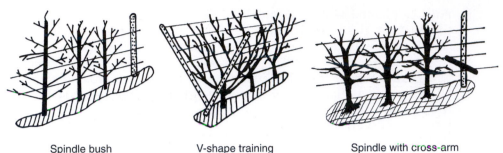

Spindle bush   V-shape training   Spindle with cross-arm

Fig. 3.1. Training of pear trees.

The slender spindle is incontrovertibly a tree form which still guarantees optimum light exposure at planting densities of 2000–3000 trees/ha. Other possibilities are V-shape training or training with a cross-arm at a height of about 100 cm, to give better development of laterals. All structural measures for particular cultivars must be designed to prevent overcrowding and shading, with the aim of optimizing production of fruiting spurs.

Choice of cultivars

Present-day cultivars originate primarily from *Pyrus pyraster* and a few other wild forms which are mainly found over an area ranging from central and southern Europe to Asia Minor. They have been cultivated for a very long time; particularly intensive breeding work was carried out in the 18th and 19th centuries in France and Belgium. Most of our present-day cultivars originate from this period. It was not until the last few years that a few important new cultivars were developed.

Whereas previously a pear orchard used to consist of a wide range of cultivars, today we have to make do with far fewer cultivars, and this is likely to continue to be the case in the future. A key factor is the type of marketing chosen by the grower. The range of cultivars is undoubtedly wider when growers have their own private marketing than in the case of cooperatives which market almost nothing else except winter pears. With private marketing, new cultivars can be added much more quickly to the product range. Figure 3.2 gives details of a selection of cultivars. The ones which are most suitable for organic production are Conference, Alexander Lukas and the new fire-blight-resistant cultivar Harrow Sweet. Of the many cultivars mentioned here, only a few, such as Williams' Bon Chrétien, Alexander Lukas and Conference, find favour for cooperative marketing. In private marketing it is important to include an early ripening variety and Doyenne du Comice in the product range, as the pears can be sold longer with this type of marketing.

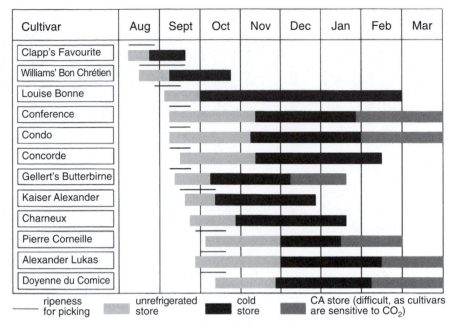

Fig. 3.2. Ripening time and storage life of important pear cultivars.

Pears, like apples, are as a rule not self-fertile. Good pollination is therefore very important. Observations have shown that Conference, for example, gives better-shaped fruit when Doyenne du Comice is present as the pollinator. Table 3.10 gives a rough guide to possible pollinators. Certainly there are a number of other cultivars which could be considered as potential pollinators, e.g. Comtesse de Paris, but most of them are difficult to market.

Table 3.10. Possible pollinators for pears.

Main cultivar	Pollinator
Conference[a]	Beurré de Gellert Doyenne du Comice[b] Williams' Bon Chrétien
Doyenne du Comice	Conference Beurré de Gellert Louise Bonne
Alexander Lukas	Conference Doyenne du Comice Williams' Bon Chrétien Charneux

[a] Partly self-fertile; [b] flowers later, but good shape of fruit.

Table 3.11. Characteristics of important pear cultivars.

Cultivar	Advantages	Disadvantages	Compatibility with quince rootstocks	Fireblight	Remarks
Précoce de Trevoux	good yields parthenocarpy little scab	fruit size soon becomes mealy	better with IS	xxx	tolerates altitude intersterile with Louise Bonne
Clapp's Favourite	fruit size moderate scab frost-hardy	vigorous fruit drop soon becomes doughy	with IS	xx	bears on the long wood
Williams' Bon Chrétien	good yields versatile use of fruit parthenocarpy	fruit drop scab *Psylla pyricola*	with IS	xx	intersterile with Trevoux and Louise Bonne
Louise Bonne	very high yield keeping quality parthenocarpy	fruit size biennial bearing scab	yes, good	xx	intersterile with Trevoux and Williams, not recommended for organic production
Kaiser Alexander	high yield not damaged by transport	stony pit scab sensitive to sulphur	only with IS	?	some tendency to biennial bearing
Concorde	high yield little russeting	similar to Conference greenish yellow after storage	yes	xx	new promising cultivar
Conference	high yield good quality hardly any scab	shape of fruit	yes, better with IS, e.g. Doyenne du Comice	xx	note selection! M202 from NL cultivar suitable for organic production
Alexander Lukas	high yields can be stored hardly any scab	flavour bacterial canker	yes	x	pressure-sensitive cultivar suitable for organic production
Doyenne du Comice	very good flavour	late, low yields susceptible to scab sensitive to sulphur	yes	xx	vineyard locations, not for organic production

Verdi	early-bearing, high yields	frost rings bacterial canker fruit thinning	only with IS	x	rather like Louise Bonne
Harrow Sweet	fireblight resistance low susceptibility to scab and *Psylla pyricola* flowers on 1-year-old wood	fruit size fruit thinning essential	yes	resistant	only fireblight-resistant cultivar that can be recommended, suitable for organic production

IS = interstock; x = moderate susceptibility; xx = medium susceptibility; xxx = high susceptibility.

Production of stone fruit (cherries and plums)

Every specialist knows that the organic production of apples requires a great deal of pioneering work, but at the present time it is impossible to tell how much effort is in fact being devoted to the organic production of stone fruit. The reason for this is that organic fruit growers have had little experience with this type of fruit and even among researchers very little attention has been paid to the organic production of stone fruit, since stone fruit has a much lower priority than apples. Interest in organic stone fruit has increased more and more, however, with the increase in the number of people buying organic produce. At the same time, one effect of this growing interest is that the consumer's quality demands with regard to organic fruit are becoming more and more 'conventional' and sales channels need to become more and more professional. This means that the producer has to try to produce fruit which both externally and internally is of high quality, in accordance with organic principles. Low susceptibility to disease and pests is a *sine qua non*, as the possibilities for combating these once they appear are very limited.

Choice of site

Since plums and sweet cherries generally flower earlier than apples, a frost-free location which is sheltered from the wind is preferable. Anti-frost irrigation is not possible for stone fruit. The soil should be well-drained. Plums tolerate a somewhat heavier soil than cherries. The choice of rootstock can compensate for these disadvantages, however.

Sweet cherries

Sweet cherries are grown successfully even in Norway. A microclimate with little risk of late frosts is more important than latitude. This occurs, for example, on sites sloping towards the north-west, west, east or south-east, where the cold air can flow away and the start of flowering is not advanced too much by temperature rises. The proximity of large bodies of water is also beneficial. The critical factor, however, is not only that the blossoms are protected from frost but also that pollination occurs. The first requirement for pollination is visits from bees, as wind pollination alone is not sufficient. Low temperatures at flowering time not only limit bee flight but also delay pollen tube growth into the ovary; this delay is often so great that a satisfactory fruit set is not achieved. It should also be borne in mind that for a normal yield the fruit set needs to be ten times higher than for apples. It is very difficult to assess a site in areas where there is very little fruit growing. In making such an assessment it may be helpful to find out whether cherry or plum trees have been grown successfully in the area before.

Cherry rootstocks

PRUNUS AVIUM (VIGOROUS). *General characteristics*:
Easy to grow, good stem builder, compatible with all sweet and sour cherry cultivars, large fruits, replant problems.

LIMBURG CHERRY. Generatively propagated standard rootstock with a straight stem.

HÜTTNERS HOCHZUCHT 170 × 53. Generatively propagated, very healthy and productive. Found mainly in northern Germany; most vigorous *P. avium* rootstock.

F 12-1. British clone, vegetatively propagated; less vigorous than the others; root suckering, poorer compatibility and productivity under marginal conditions; still important mainly because of its easy propagation.

CHARGER, PONTARIS, PONTAVIUM. Experimental rootstocks.

PRUNUS MAHALEB (SEMI-VIGOROUS TO VIGOROUS). *General characteristics*:
Only for light soils, otherwise some trees die after 6–10 years, including sour cherries. Highly susceptible to damage by mice, little root suckering, frost-hardy.

SAINT-LUCIE 64 (VEGETATIVE). 80–90% of the vigour of Limburg cherry on a suitable site; the most important cherry rootstock in France.

HEIMANN. Best-known clone in Germany.

PRUNUS CERASUS (DWARFING TO SEMI-VIGOROUS). *General characteristics*:
Curbs growth, frost-hardy, not always compatible, medium to high soil requirements, good fruit size can be obtained by pruning and additional irrigation. Not suitable for replanting after *P. avium* or after itself.

TABEL = EDABRIZ

Origin:	Iran, selected in France
Compatibility:	compatible with all cultivars up to now (according to the literature); only slight burr knot formation
Cropping performance:	very good
Start of fruiting:	very early
Dwarfing:	about 70% of Limburg cherry, probably too weak for organic production
Other aspects:	good nursery performance, susceptible to chlorosis, stake

Recommendation: on a trial basis for dense plantings; in areas with lower rainfall only on soils with a soil score of more than 70 without trickle irrigation; for vigorous cultivars

WEIROOT CLONES. Early and high yields (from third year after planting onwards), incompatibilities – if not absolutely virus-free – with Van, Sam, Charmes and various Swiss cultivars, high soil requirements.

- W72, W53 (70–80% growth reduction relative to Limburg cherry)
- W158 (50% growth reduction relative to Limburg cherry)
- W154 (40% growth reduction relative to Limburg cherry)
- W13 (25% growth reduction relative to Limburg cherry)

Hybrids

COLT
Cross: *P. mahaleb* × *P. pseudocerasus*, UK
Compatibility: medium, frequent aerial roots, burr knot formation
Cropping performance: good
Start of fruiting: early
Dwarfing: 10–30% weaker than Limburg cherry
Other aspects: shallow root system yet good anchoring, tolerates winter frost
Recommendation: instead of Limburg cherry on good, moist soils, good development in the juvenile phase; very good in the nursery; not on dry sites, otherwise risk of small fruit; suitable for replanting

MAXMA DELBARD 14
Cross: *P. mahaleb* × *P. avium*, Oregon, USA
Compatibility: good, little or no burr knot
Cropping performance: good
Start of fruiting: early
Dwarfing: 30–40% weaker than Limburg cherry
Other aspects: hardly any chlorosis, moderate *Phytophthora* susceptibility, ramification somewhat weaker, deep-rooting
Recommendation: grow on a trial basis, as experience in France has been very good; grown there in a wide variety of soils and climates; plant on lighter soils rather than on very heavy or waterlogged soils

GI-SEL-A 5 (PREVIOUSLY KNOWN AS 148/2)
Cross: *Prunus cerasus* × *Prunus canescens*
Compatibility: good up to now, if material virus-free

Cropping performance:	early cropping, tendency to overcropping in fertile cultivars from the third year after planting
Dwarfing:	50–60% weaker than Limburg cherry
Other aspects:	flat and uniform ramification; difficult to propagate, only via tissue culture up to now; sensitive to virus-affected material, then defoliation
Recommendation:	particularly suitable for vigorous, weak carriers, such as Summit or Regina; without trickle irrigation only on good soils with these cultivars; on traditional cherry sites winter pruning from the fifth–sixth year after planting to stimulate regrowth

Cherry cultivars

As long as there is no adequate possibility of controlling the cherry fruit fly in organic production, the choice of cultivars is limited to those ripening in the first–fourth week. There may also be sites where infestation occurs earlier, however. Susceptibility to cracking and subsequently susceptibility to *Monilia* are also critical factors in any recommendation. Susceptibility to shot hole and bitter rot should also be taken into account in choosing a cultivar. Unfortunately many cultivars that are resistant to cracking are soft-fleshed. Except for Geisepitter, all the cultivars given in Table 3.12 are dark-red sweet cherries.

Plums

Plum rootstocks

MYROBALAN (VIGOROUS).   *General characteristics*:
Good compatibility, can be propagated generatively, suitable for dry and light soils, replanting and heavy-cropping cultivars; good anchorage.

Selections with improved fertility:	Myruni, Hamyra
Cropping performance:	starts late
Root suckering:	little
Susceptibility to plum pox:	low
Drought:	good resistance
High pH:	suitable
Waterlogging:	medium to good

ST JULIEN ROOTSTOCKS (*P. INSITITIA*).   *General characteristics*:
Vigour 70–80%, ramification flat and good, tolerant to plum pox but quite a lot of root suckering; generatively propagated down to INRA 2; good anchorage.

Table 3.12. Cherry cultivars.

Cultivar	Origin	Week of ripening	Flowering	Growth	Start of cropping	Susceptibility to cracking	Susceptibility to rot	Remarks
Early Meckenheim	Pfalz, Germany	2nd	early	semi-vigorous, upright	early	low	low	fairly soft, medium-size
Moreau (S. des Charmes)	France	2nd	early	vigorous, lateral	late	medium	medium	first promising market cultivar, ripening spread over a period
Merchant	UK (FA 'Merton Glory')	3rd	moderately early	semi-vigorous, good branching	medium	low	low	good market cultivar, medium-size
Johanna	Jork, Germany	3rd	moderately early	semi-vigorous, semi-upright	medium	low	low	tends to have small fruit but very resistant to cracking
Geisepitter	Middle Rhine, Germany	3rd	moderately early	vigorous, pendulous laterals	early	low	low	cultivar with light-coloured fruit, soft, impact bruises
Giorgia	Italy	3rd–4th	moderately early	axial	early	relatively high	medium	too susceptible to cracking in a damp climate; very firm
Starking Hardy Giant	Wisconsin, USA	4th	medium	semi-vigorous, broad	early	medium	medium	firm, medium-sized fruit; virus-free material essential

Selections:	GF 655/2, St Julien A
Cropping performance:	starts fairly early
Root suckering:	a lot; depends on cultivar and site, St Julien A rather less than GF 655/2
Susceptibility to plum pox:	low
Drought:	relatively good resistance, GF 655/2 rather better than St Julien A
High pH:	suitable
Waterlogging:	also suitable for heavy soils

GF 8/1 (*P. CERASIFERA* × *P. MUNSONIANA*).   *General characteristics*:
Vigorous (100%), but considerably more fertile than Myrobalan, fairly resistant to nematodes and root rot, susceptible to plum pox.

Cropping performance:	starts early
Root suckering:	quite a lot of root suckers which appear late in growth
Susceptibility to plum pox:	high, re-infection via shoots in autumn
Drought:	relatively good resistance
High pH:	suitable
Waterlogging:	good resistance

FERELEY (= 'JASPI®') ('METHLEY' × *P. SPINOSA*).   *General characteristics*:
Vigour comparable with St Julien, i.e. semi-vigorous; higher initial fertility and hardly any root suckering; because of micro-propagation, suckers can develop from the base of the stem; suitable for low-yielding cultivars, such as Hauszwetschke or Mirabelle.

Cropping performance:	starts early
Root suckering:	few root suckers, growing underground from the base of the stem
Susceptibility to plum pox:	low up to now
Drought:	good resistance up to now
High pH:	suitable
Waterlogging:	very good tolerance (selected for this in France)

ISHTARA ((*P. CERASIFERA*) × (*P. PERSICA* × *P. BELSIAN*)).   *General characteristics*:
Semi-vigorous, heavy-bearing rootstock with very early cropping, but with good fruit size; brittle root system, somewhat steep crotch angle; high soil requirements; age potential of the rootstock not yet established; not for heavy croppers, but for late-cropping, small-fruited cultivars such as Nancy Mirabelle or Hauszwetschke.

Cropping performance:	very early
Root suckering:	none
Susceptibility to plum pox:	low up to now
Drought:	only suitable under certain conditions
High pH:	not suitable above pH 7.0
Waterlogging:	not suitable

Plum cultivars

A major problem of plum cultivation in Europe is the viral disease plum pox, which cannot be directly controlled. It is not widespread in northern Europe but determines the choice of cultivars in southern and south-western Europe. Damsons are generally preferred in Germany and eastern Europe, whereas plums and greengages are more important in other countries. Professional advice should always be sought when choosing a cultivar for a particular locality.

Organic production of small fruit

Organic small fruit are much in demand. That does not mean, however, that all organic small fruit finds a buyer. Only growers who get marketing under control at an early stage and offer top quality are guaranteed a market. In addition to good specialist knowledge and a lot of flair, the growing of organic small fruit requires continuous training to keep abreast of developments. The growing of strawberries, in particular, is changing rapidly in terms of cultural techniques, but bush fruits, too, are being affected by widespread changes, such as the increasing use of rain shields to provide protection from adverse weather.

Choice of site

Airy sites in full sun, but protected from strong winds and late frosts, are the most suitable sites for all species of small fruit.

Soil quality

The quality of the soil is of particular importance. Spade samples should always be taken to get some idea of the soil before any new planting is carried out.

Points to bear in mind when taking a spade sample are:

- use a drainage spade if available
- take the sample at a depth of at least 40 cm
- rust spots and black manganese concretions indicate air circulation problems and thus unsuitable soils.

STRAWBERRIES AND RASPBERRIES. These are the most demanding in terms of soil requirements. They should be grown only on medium-heavy to light, free-draining soils. Areas with compaction or waterlogging are unsuitable. Reduced yields and root disease problems are unavoidable on unfavourable soils. Planting by the hill system is to be recommended if soil conditions are not entirely optimal.

Table 3.13. Plum cultivars for organic production (selection).

Cultivar	Ripening time	Fruit	Tree	Plum pox	Remarks
Herman	1st–2nd week	round, dark blue, sweet, 35 g	dwarfing, broad	tolerant	pre-harvest fruit drop, very good for baking, better than R. Gerstetter
St. Hubertus	2nd–3rd	roundish, blue, yellowish flesh	semi-vigorous, broad	tolerant	only in early locations ripens sooner than Ersinger, not freestone every year
Katinka	2nd–3rd	oblong, colouring like Hauszwetschke, 32 g	steep, dwarfing	susceptible if attack is severe	pure baking plum, very high yield, plant close together, new cultivar
Ersinger	3rd	oval, sharp flavour, skin reddish	semi-vigorous, flat-branching	only leaf symptoms	fresh consumption and baking, thin sometimes, quickly goes soft, weak point: colouring
Cacaks Schöne	4th	oval, blackish blue, flesh greenish, 40 g	steep, dwarfing	leaf symptoms possible	baking plum for caterers, very high yield, big demand in the trade
Julia	4th	flattened oval shape, colour like Hauszwetschke, tart–aromatic	vigorous, steep	only leaf symptoms, few fruit symptoms	late cropping, appreciated for baking, also fresh consumption
Auerbacher	6th	roundish – drop-shaped, not always blue, flesh yellow, 32 g	semi-vigorous, broad, often bare inside	highly susceptible	baking plum much in demand in the trade, susceptible to Valsa, so only suitable for light soils, tends to have small fruit
Hanita	6th	oblong, dark blue, 40 g	steep, little branching, semi-vigorous	tolerant	uniform yield, baking still possible, not always freestone, grow where Auerbacher does not thrive
Ortenauer	7th	oblong, light colour, 32 g	semi-vigorous, flat-branching	highly susceptible	baking plum, heavy-bearing, stores well
Cacaks Fruchtbare	7th	oval, blue, 35 g	steep, dwarfing	susceptible if attack is severe	fox-red when tree is overcropped, good flavour, medium baking quality
Nancy-Mirabelle	6th–7th	round, yellow	broad, vigorous, little branching	highly tolerant	brandy, baking, fresh market, best type 1510
Hauszwetschke 1. Meschenmoser 2. Schüfer 3. Etscheid or Purpurgold	7th–8th	oval, sharp sweet flavour, 32 g	very vigorous, broad	highly susceptible	best cultivar for baking, brandy, fresh market; choose rootstocks that give early cropping (Ferely or Ishtara)
President	8th	oval, large, reddish, insipid	semi-vigorous, very steep	only leaf symptoms, few fruit symptoms	fresh market, best cultivar for storage
Elena	9th–10th	wide drop shape, flesh greenish, 35 g	semi-vigorous, good branching	tolerant	new late-ripening Hohenheim cultivar with sharp flavour and good baking quality; rather irregular fruit size

BRAMBLES, CURRANTS AND GOOSEBERRIES. These are less demanding than raspberries.

BLUEBERRIES. Thrive only on light acid soils with low salinity. Depending on the author and the method of measurement, the ideal pH of the soil ranges from 3.5 to 5.5. In many sites special treatment is needed to achieve this.

Altitude

- Frost-tender cultivars are not suitable for growing at altitude.
- Bush fruits must be protected from the pressure of snow by training systems of very stable construction.
- The harvest is delayed by 3–5 days for every 100 m of increasing altitude.

Strawberries: can be grown up to an altitude of about 1600 m above sea level.
Raspberries, currants and gooseberries: can be grown up to about 1400 m above sea level.
Brambles and blueberries: can be grown up to about 1000 m above sea level.

Previous crops

As a general rule, the less closely related the previous crop and the berry species envisaged, the smaller is the risk of transfer of harmful organisms.

Strawberries

After strawberries no other crops should be grown for at least 3–4 years or, if there has been infection with red root rot (*Phytophthora fragariae*), at least 15 years. To allow sufficient time for careful preparation of the bed where the strawberries are to be planted, the previous crop should be removed 4–6 weeks before planting. Suitable previous crops are 1-year artificial meadows, oil radish, rapeseed, mustard, buckwheat, winter cereals or vegetables (except *Fabaceae*). Previous crops that are unsuitable are natural meadows which have been turned over less than 3 years ago (because of weeds, white grubs, wireworms and nematodes), potatoes and tomatoes (because of *Rhizoctonia* and *Verticillium* sp.), phacelia (because of rhizome rot) and weed-infested crops.

Bush berries

Growing an intermediate crop for 1 or 2 years, e.g. oil radish, an artificial meadow with lucerne or another deep-rooting sown plant, improves the structure of the lower soil layers. When bush berries are grown after a natural meadow there is an increased risk of weed invasion and wireworm infestation.

Fertilizers

Application of fertilizers is based on analysis of the soil (taking into account the vigour of the previous crop in the case of new plantings). Soil analyses, which provide information mainly about nitrogen levels, should be carried out at least once every 5 years. To guarantee comparability of the results, it is best to have the analysis always done by the same laboratory.

Nitrogen (N) supply

Only organic fertilizers can be used in organic farming. To make sure that the nitrogen is available to the plants at the time of greatest need, it is important to take into account the time needed to convert the nitrogen to a form that is available to the plant (speed of action). The speed of action depends on the fertilizer, the soil conditions and the weather. Farm manures can also be used, except for blueberries. Only 10% of the total nitrogen in compost and 50% of that in cattle manure can be included in the nutrient balance in the first year. In the case of other nutrients, 100% is included.

STRAWBERRIES.　Undiluted slurry or fresh manure should already be added to the soil under the previous crop, so as to avoid scorching of the strawberry plants. Nitrogen requirements are particularly high 2–3 weeks after planting. Optimum use of fertilizer releases sufficient nitrogen at this time. Hoeing can also increase mineralization and availability of nitrogen to the plants during this period. Blood and horn meal or fast-acting commercial preparations can be used if additional nitrogen fertilizers are needed (see *Permitted fertilizers and soil conditioners (Annex II to EU Regulation 2092/91)* in the Appendix).

　　N.B. Excessive amounts of nitrogen encourage harmful organisms, reduce yield and fruit quality and pollute groundwater, as well as causing unnecessary costs.

Phosphorus, potassium and magnesium supply

Sufficient quantities of phosphorus (P), potassium (K) and magnesium (Mg) are usually added to the soil along with organic material (compost or manure). The deliberate addition of these minerals is only advisable if there is some evidence of deficiency or if supply levels are in categories A, B or C (low, moderate or sufficient).

Planting stock

Strawberries

GREEN PLANTS.   Green plants are obtainable in either potted or bare-rooted form. Bare-rooted plants are more heat-sensitive and have higher water requirements than potted plants. Since they have to be planted 7–10 days earlier than potted plants, they are often not available at the right time. As a result, they are not very important.

REFRIGERATED PLANTS.   Refrigerated plants are lifted during dormancy and stored at −1.5 to −2°C. On the day before planting they are thawed at 4–8°C and then planted immediately. This type of plant is usually sold bare-rooted.

Refrigerated plants are more susceptible to harmful organisms. For this reason, as a general rule, green plants are preferable to refrigerated plants.

Bush fruits

Raspberries: green plants (potted, not woody)
Brambles: potted plants
Currants and gooseberries: bare-rooted outdoor plants (1–4 shoots, depending on training system)
Blueberries: plant with soil ball, 3-year-old, 40–60 cm high container plant, 2-year-old, 30–40 cm high

Choice of cultivar

The most important criteria in choosing a cultivar are high internal fruit quality, high resistance to harmful organisms and highly reliable yields. Since cultivars for small fruits, especially strawberries, are continuously changing, the information below must be treated with due caution. Moreover, cultivars do not exhibit the same performance at all sites, so there may well be variations from the information given.

Strawberries

Wädenswil 6: early, highly aromatic cultivar, but poor transportability makes it suitable only for domestic gardens, pick-your-own and direct marketing.
Darline: early, aromatic cultivar, light-coloured and firm fruits; loose plant structure and rather low yields (reduce distance between plants).
Thuriga: medium ripening time, highly aromatic, large and firm fruits. Rather susceptible to leaf spot.

Majorat: medium ripening time, aromatic, firm fruits. Popular cultivar among Swiss organic growers in spite of the somewhat increased susceptibility to mildew and leaf spot.

Raspberries

Meeker: medium ripening time, medium-sized, aromatic fruits that are easy to transport. Highly vigorous cultivar with increased frost tenderness. Not very susceptible to root rot.
Nootka: medium ripening time, highly aromatic, medium-sized fruits, only moderate transportability. Not very susceptible to root rot.
Rubaca: medium ripening time, medium-sized, aromatic fruits, only moderate transportability. Extremely resistant to root rot.
Autumn Bliss: autumn-bearing cultivar, large, moderately aromatic, transportable fruits; extremely resistant to root rot.

Brambles

Theodor Reimers: early, small-fruited, highly aromatic cultivar, poor transportability.
Loch Ness: early, thornless, large-fruited and aromatic cultivar, very good transportability; increased susceptibility to false mildew in Switzerland.
Navaho: medium ripening time; thornless cultivar with upright growth habit and large, aromatic and easily transportable fruits.

Redcurrants

Jonkheer van Tets: very early, increased tendency to discoloration.
Rolan: medium ripening time, rather sensitive to rain.
Rovada: late, widely distributed cultivar, very robust.

Blackcurrants

Bona, Ceres and Titania: generally robust cultivars, but only moderately suitable for hand picking.

Gooseberries

Invicta: green, robust cultivar with large fruits.
Rokula: red, robust cultivar with medium-sized fruits.
Hinnonmäki yellow: yellow, robust cultivar with medium-sized fruits.

Blueberries

Duke: early ripening time.
Bluecrop: medium ripening time with long harvesting period, standard cultivar until the present time.
Coville: late ripening time.

Planting times

Strawberries

In normal cultural practice with green plants, the planting time is between the end of July and the middle of August. Weak-flowering, large-fruited cultivars should be planted before heavy-flowering, small-fruited cultivars. Late planting times result in less stooling and thus a slightly earlier harvest. If planting is carried out after mid-August, the strawberries need to be planted closer together in order to compensate for the lower yields per plant.

Bush fruits

As a rule, planting must be carried out in soil that is not frozen and not too wet.

Raspberries: green plants (potted, not woody), 20 May to end of June.
Brambles: potted plants, end of March to end of June.
Currants and gooseberries: bare-rooted outdoor plants, start of November to end of April.
Blueberries: container plants, start of September to end of October; plants with soil ball, start of November to end of April.

Plant spacing

Strawberries

Plant spacing depends on the cultural system, mechanization and the growth properties of the cultivars. Good air circulation through the crop reduces attack by harmful organisms such as grey mould. For this reason, when the foliage has fully developed, there should be only slight contact between neighbouring plants within the row. In normal cultural practice a suitable spacing is 80–110 cm between the rows and 25–50 cm within the rows. If the plants stool poorly (this depends on the cultivar), the yield per unit of area can be increased by putting one stolon per plant into the row and tending it in the same way as the parent plant.

Bush fruits

Raspberries: spacing 2.50 m between rows, 0.40–0.70 m between plants.
Brambles – horizontally growing cultivar: spacing 2.50–3 m between rows, 3–4 m between plants.
Brambles – vertically growing cultivar: spacing 2.50–3 m between rows, 1.60–2 m between plants.

Currants and gooseberries: spacing 2.50–3 m between rows, 0.60–1.20 m between plants.

Blueberries: spacing 2.50–3 m between rows, 1–1.50 m between plants.

Planting technique

Strawberries

If the area to be planted is more than about 0.1 ha it is worth using a planting machine.

N.B. Never plant strawberries with dry pot balls or roots. To make sure that the plants grow as well as possible, the roots of young plants that are not in pots should not be bent in the soil. If the roots are too long, they can be shortened slightly.

For **bush fruits**, planting is usually by hand.

HOW LONG SHOULD STRAWBERRY PLANTS BE KEPT? Strawberries are normally grown for 1 year or for 2 years. When organic production methods are used the **1-year system** has the following advantages over the 2-year system:

- fewer problems with root diseases, grey mould, leaf spot and strawberry mildew
- bigger fruits which separate better from the calyx, and thus greater harvesting efficiency
- lower costs of weed control
- earlier cropping.

The **2-year system**, on the other hand, is useful in the following cases:

- growing strawberries at altitudes over 1000 m
- varieties with weak flowering and large fruits
- pick-your-own.

POST-HARVEST CARE OF 2-YEAR CROPS. In dense plantings, every second plant should be removed, or the strawberries should be planted further apart at the outset. After the harvest, the leaves should be mown off with a rotary mower or mulching flail cutter, before the weeds release their seed. If a mulching flail cutter is used, the leaves can be left lying on the ground, otherwise they should be removed.

N.B. The growing tips (hearts) of the strawberry plants must be left intact. Remove plants that are severely infested with strawberry mites. Top dressing is applied if necessary. Subsequent care of the plants is the same as for freshly planted strawberries.

Training systems for bush fruits

Choosing the appropriate training system is a key factor for the health of the crop and for yield and harvesting efficiency.

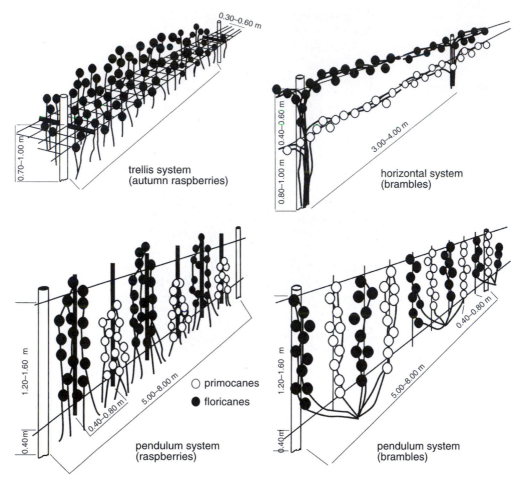

trellis system
(autumn raspberries)

0.30–0.60 m

0.70–1.00 m

horizontal system
(brambles)

0.40–0.60 m

0.80–1.00 m

3.00–4.00 m

pendulum system
(raspberries)

1.20–1.60 m

0.40 m

0.40–0.80 m

5.00–8.00 m

○ primocanes
● floricanes

pendulum system
(brambles)

1.20–1.60 m

0.40 m

0.40–0.80 m

5.00–8.00 m

Fig. 3.3.

Summer raspberries

Summer raspberries can be trained both by the **hedgerow** system and by the **pendulum** system.

Autumn raspberries

Unlike summer raspberries, autumn raspberries fruit on current-year canes. The **trellis** system is the simplest training system.

Brambles

The training system depends on the growth characteristics of the cultivar. Horizontally growing cultivars are trained by the **horizontal** system,

Fig. 3.4. Hedgerow training of currants.

whereas cultivars with upright growth are trained vertically by the **pendulum** system. Layering the canes (to protect them from frost) is only possible with horizontally trained cultivars.

Currants and gooseberries

Whitecurrants and redcurrants are trained by the **hedgerow** system (see Fig. 3.4). There should be a distance of 0.30–0.40 m between leaders within the row. Air circulation and drying is much better in the hedgerow system than when currants are grown as bushes. Vigorous 1-year-old side-shoots from the leaders give the best fruit quality. A stump of 2–3 cm is left when the side-shoots that have fruited are pruned. Good fruiting canes usually develop from this. Leaders should be renewed after 3–4 years.

Blackcurrants

Blackcurrants are generally trained as a **bush** (an advantage in mechanical harvesting). The bushes should be planted deep enough for adequate production of shoots from the base. The advantage of training as bushes is the low initial costs; the disadvantage is the low efficiency of harvesting by hand. To improve air circulation, only 6–8 fruiting canes should be left per bush; berries should not be picked from the fruiting canes for longer than 3 years.

Blueberries

Blueberries are trained as **bushes**. Whole fruiting canes should be removed if fruit size is inadequate or planting is too dense.

Weed control

Strawberries

The previous crop has a significant effect on weed infestation of strawberry plots. For weed control the soil is either left bare or covered.

BARE SOIL.   Weeds are removed as soon as possible by hoeing and/or flaming between the rows and by hoeing by hand within the rows. Runners are also cut out in the same operation. Excessively vigorous hoeing destroys the soil structure and causes problems with slaking and erosion. What is important is that the plot should be free of weeds at the start of the winter, as many weeds and grasses continue to grow even at low temperature and can achieve dense growth by the spring.

COVERING THE SOIL.   Covering the soil with mulch films or organic material is an alternative to leaving the soil bare. In high-rainfall areas with heavy soil, however, this system tends to lead to waterlogging.

Bush fruits

All bush fruit species, except for vigorous bramble cultivars, show a sensitive reaction to companion plants in the strip. Deliberate use of ground cover in the strip, however, can counteract excessive shoot growth or delayed cessation of shoot growth and thus have a positive effect on the physiological balance of the plants. Ground cover of this type binds nitrogen during the winter and makes this nitrogen available to the plants again after hoeing in the spring. Under-vine sweepers such as used in vineyards can be used for hoeing, but care must be taken to avoid damaging canes (creating entry points for harmful organisms). Particular care is needed when hoeing among raspberry plants, as hoeing makes them push their roots into deeper soil layers. This increases the danger of root rot.

Using straw to keep strawberries clean

A layer of straw is placed beneath the strawberry plants to make sure that the fruits stay clean and to prevent the spread of grey rot and rhizome rot (*Phytophthora cactorum*). In addition the straw conserves soil moisture (this can also be a disadvantage), suppresses weeds and makes it easier to move around. The straw (80–100 kg/acre) is put down just before the fruits touch the soil. If this job is carried out before or during flowering, there is an increased risk of damage from late frost (because the soil is heated up less and radiates less heat during the night). In order to reduce the amount of straw to a minimum, only well-threshed wheat straw should be used. As an alternative to wheat straw it is possible to use elephant grass, which is free of weed seeds, but it is not yet clear whether it is suitable for strawberry cultivation. On smaller plots the straw is put down by hand, but on bigger fields machinery can be used.

N.B. Straw which is too finely chopped sticks to the berries in wet weather.

Irrigation

Water requirements are highest when the fruits are being formed; an adequate water supply during this period has direct effects on fruit size and yield. In most areas, irrigation is absolutely essential when raspberries are grown on raised beds or when blueberries are grown in a substrate. If soil temperatures in the spring and autumn are low, to counteract root rot raspberries should be watered only in exceptional cases. Blueberries should preferably be watered with soft water, e.g. collected rainwater. Irrigation makes it easier for currants and gooseberries to attain their optimum canopy height. Trickle irrigation is preferable to spray irrigation for bush fruits, as it saves water, the fruits stay dry and the plot remains accessible to vehicles.

Fig. 3.5. Covering bush fruits – an important preventive measure against botrytis.

Bees

Bees are important for fruit quality in strawberries, raspberries and brambles and for yield in currants and gooseberries. Four to six vigorous bee colonies per hectare are usually sufficient.

Rain shelter for prevention of diseases

Under rain shelters, bush fruits are at least partially protected from various diseases (e.g. grey rot) (see Fig. 3.5). In addition, plants that are covered by a shelter produce fruits with better keeping quality, and the time of picking, as in the case of currants for example, can be postponed to some extent. The disadvantages of rain shelters are the very high installation costs, the relatively high consumption of energy and resources in manufacture, and the adverse effect on the landscape.

Application techniques

Unlike pome and stone fruits, there is still a need for a great deal of research into application techniques for berry fruit. Spraying is carried out with knapsack sprayers or spray guns on smaller plots, and with air-jet sprayers on larger plots.

The plants can be damaged if the spray pressure is too high or the air currents too strong. The amounts of spray recommended at present vary widely, depending on the stage of development of the crop, the harmful organisms to be controlled and the selected application device. They can range from 200 l/ha in the year of planting to 2000 l/ha when the foliage is fully developed. The effectiveness of application can be gauged by placing water-sensitive paper at various locations among the plants. In order to reduce soil compaction to a minimum, use of wheeled machinery should be limited to times when the soil is dry, and the machinery should be of a type which exerts little pressure on the soil.

4 Cultural measures in organic fruit growing

Protection of the soil when using machinery

In organic fruit production, particular care should be taken to use machinery in a way that does not damage the soil. The larger pores in the soil are responsible for aeration and water storage and also serve as habitats for soil organisms. It is mainly the larger pores which are reduced in size by pressure on the soil. When planning the use of machinery in such a way as to conserve the soil it is important not only to **prevent compaction** from the pressure exerted by vehicles on the soil but also to **avoid tearing up the sward** in the tramlines of orchards with permanent green cover. Soil conservation is affected by the type of tractor, the tractor tyres, and the number of times vehicles are driven through the orchard.

Tractor design

The degree of damage caused to the soil by a tractor depends on the weight, which is distributed over the area covered by the tyres. It should be borne in mind that in spite of an equal specific load on the soil (contact area pressure) a heavier tractor (even if it has bigger tyres) will exert greater pressure on the soil.

The basic rule is: drive into the orchard **as little as possible, and with as light a tractor as possible**. Only a light tractor weighing less than 2000 kg, and with broad tyres (at least 35 cm), conserves the larger pores of the soil.

In terms of construction, the most suitable tractors are articulated ones which in addition to a low total weight have a favourable weight distribution (60% on the front axle and 40% on the rear axle). This type of construction makes it possible to use 40-cm-wide tyres of the same size on the front and rear wheels. At the same time it is possible to keep the external width of the tractor below the maximum of 135–140 cm required by modern organic orchards. A further advantage of this type of construction is the good manoeuvrability and high adaptability to different types of terrain.

It should be borne in mind that articulated tractors are not suitable for heavy traction and transport work. This must be done with standard (compact) tractors.

Before buying a tractor, therefore, it is important to decide which type fits better into the development plan of the organic enterprise: the **versatile compact tractor** or the **soil-conserving articulated tractor** as a second tractor for the orchard.

Tyres

Apart from their suspension characteristics, the principal function of the tractor tyres is to transfer propulsive and braking forces to the ground without damaging the sward. On level ground it is much easier to use machinery in a way that conserves the soil than in orchards on sloping ground, where the safety of the driver becomes of paramount importance as the gradient increases. On sloping ground it is not possible to achieve the optimum in terms of both requirements (transfer of propulsive and braking forces and conservation of the sward) with a single type of tyre. In this case, when choosing tyres, compromises have to be made in favour of the propulsive and braking forces, so as to be able to drive the tractor into the orchards for plant protection work even after rain has fallen.

The following types of tyres are suitable for orchard tractors.

Radial tyres

Radial tyres of the grassland type are particularly suitable for transferring propulsive and braking forces. Compared with arable tyres, the tyre tread is characterized by a lower cleat height and closer cleat spacing. These grassland tyres, which are also called universal tyres, have the advantage of getting a sufficient 'grip' or purchase on the slope and thus guaranteeing adequate safety, although they have the disadvantage of sometimes damaging the sward.

Wide tyres

Both Terra tyres and twin tyres are called wide tyres.

Terra tyres are low-pressure wide tyres of low cross-section for ground that cannot bear a heavy load. They are very soft tyres which adapt extremely well to the ground surface and do very little damage to the sward. If the grass cover is unbroken, this type of tyre can also be used on sloping ground. A short section of bare soil, however, makes this tractor unmanoeuvrable. In addition, the soft tyres are regularly damaged by prunings.

Twin tyres are wide tyres of low cross-section for normal use and can be chosen in rubber mixtures of various degrees of hardness with a high-purchase tread or a tread that does not damage the sward. These wide tyres usually need special rims.

Tyre pressure

The tyre pressure is also critical for conservation of the soil. With a lower tyre pressure, the area covered by the tyre increases while at the same time the specific pressure exerted on the soil (kg/cm^2) is reduced. As a general rule, tyres with as large a volume as possible should be chosen, so as to be able to keep the tyre pressure as low as possible (<1 bar), taking into account the minimum load-bearing capacity. The front wheel tyres of compact tractors, in particular, are usually too small, with the result that the area covered by the tyre is too small and excessive damage is done to the soil.

Number of times vehicles are driven through the orchard

The number of times vehicles are driven through the orchard, and thus the damage done to the soil, depends to a large extent on the choice of the **pesticide spraying method** and the **method of harvesting**. Of the two methods of pesticide application – normal spraying (using 1000–1500 litres of water/ha) and low-volume spraying (using 150–300 litres of water/ha) – the latter is more efficient and does less damage to the soil. Apart from the fact that the distance driven in the orchard is less, because less frequent trips are made to fill the tank, other considerations which make low-volume spraying preferable for organic fruit growers are the lower volume of water carried and the very short time needed for optimum pesticide application in the morning or evening. It is very important, however, to choose suitable nozzles and filters, and to adjust the equipment properly (adjustment of the spray distribution to the foliage of the trees and adjustment of the air flow rate to the penetration resistance of the trees).

The effect of the method of harvesting on soil conservation is primarily determined by the transport of pallet boxes. Picking into pallet boxes on harvesting trucks greatly reduces the number of times vehicles are driven into the orchard, and thus the damage to the soil, compared with the transport of fully loaded pallet boxes with the tractor stacker from the orchard.

Care of the soil

Objectives of soil care measures in organic fruit production

- To support the performance of the tree (vegetative and generative) and quality formation.
- To create or maintain a healthy soil structure and soil activity at a high level.

Trees planted with dwarfing rootstocks and with closely spaced planting have a small root system, mostly near the surface, which is correspondingly sensitive to competition for water and nutrients. Since herbicides and readily soluble, synthetic fertilizers are not allowed, control of competition from companion plants must be achieved by other methods. Apart from hoeing, which is the most widely used method of soil care in organic orchards, the rows of trees can also be covered with organic materials such as **bark mulches** or **rapeseed straw**. Robust synthetic films have been found to be effective in some organic orchards. All herbicides are prohibited in organic fruit production – even if they also include natural substances.

Basic rules for organic care of the soil

1. Optimum protection of the soil is only guaranteed if the soil is protected by a ground cover crop. The fruit grower needs to decide when exceptions need to be made to this rule. The competition to the fruit tree from the ground cover crop is highly dependent on vegetation and varies widely, depending on climate, soil and tree form.

2. The removal of the ground cover in the row of trees depends on the degree of competition at the time. The specific situation must be taken into account in deciding what needs to be done, and when.

- Tests to determine the **competition for water**: sensors, tensiometers, etc.
- Tests to determine the **competition for nitrogen**: N_{min} soil tests in stage C (greentip-mouse-ear stage), leaf analyses in the T stage, optical assessment.

Nitrogen requirements are highest for the fruit tree in the spring after flowering until the T stage of the fruits. In this phase the competition from a green cover crop is particularly severe, as the companion plants need the most nitrogen for themselves. The same is true of summer drought. Here, too, the roots of the grasses and other plants can compete better than the fruit tree roots. For this reason the planting strip often needs to be kept free of weeds in these periods that are critical for the development of the tree and fruit. In autumn – a period with low nitrogen requirements – a ground cover crop on the planting strip is associated with a number of positive effects, especially on fruit quality:

- improved colour development of the fruits
- better keeping quality (less nitrogen and potassium in the fruits)
- reduction of oversize
- prevention of nitrogen leaching
- the entire area of ground should be of the highest possible ecological quality.

The rows of trees and tramlines should be tended in such a way as to create good conditions for beneficials and species diversity. Mowing does less harm to beneficials than mulching. The tramlines and rows should be mown alternately.

Wild-flower meadows can only be established after sowing.

Soil care by the sandwich system

The 'sandwich' system is a method of soil care which was developed at the Research Institute for Organic Farming, Frick, Switzerland. It has four main aims:

- use of simple and therefore cheap implements without a feeler arm
- maintenance of the proven beneficial effect of companion plants on the soil and indirectly also on fruit quality
- avoidance of damage to trunks and roots
- avoidance of or reduction in manual labour in the area around the trunks.

The principle of the sandwich system is that a strip about 50 cm wide on either side of the row of trees, outside the area of the trunks, is hoed, while at the same time low ground cover is grown on a 35–50 cm wide strip in the middle of the row, without mulching. The total volume of soil without competition from vegetation, per tree, thus remains just as large as if the whole area were hoed, or it can be varied as required.

Equipment for undertree soil care

TREE ROW MULCHER.   Any fast-response mulcher (e.g. electrohydraulically controlled) can be used as a tree row mulcher. With increasing planting density, however, there are difficulties with speed of movement and problems with trunk damage.

INTER-ROW SWEEPER.   The number of scab spores can be reduced by sweeping out the leaves and shredding them. Furthermore, the ground which has been swept will be capable of giving off more heat in the night during late frosts, allowing the temperature to increase by a few degrees.

Inter-row sweepers should preferably be mounted on the front of the tractor and combined with a prunings shredder.

MECHANICAL TREE ROW EQUIPMENT.   Three types of machines are available for controlling grass growth.

Table 4.1. Comparison of various alternative methods of soil care in organic fruit production (pome fruit), as recommended for commercial fruit growers by the Federal Research Institute for Fruit Growing, Viticulture and Horticulture, Wädenswil, Switzerland.

Method	Remarks	Where suitable?
ground cover in rows	Ground cover is beneficial for soil structure and prevents erosion and nutrient losses. Positive effect on fruit quality. Sown ground cover is very labour-intensive and does not last long enough, so a natural ground cover is recommended. The row can be tilled mechanically in the spring, when the trees have high nitrogen requirements (April, May), or in summer, if there is drought, so that the competition from weeds is temporarily eliminated. Mulching or mowing around the base of the trunk is not entirely satisfactory with the equipment currently available on the market. The vegetation must be cut lower over the winter, and regular, careful checks must be made for mice.	From about the 4th year after planting in sites with adequate growth and sufficient rainfall
winter ground cover in rows	As for permanent ground cover, but mechanical tilling of the row in the spring (before the trees flower) and if necessary in summer, in order to reduce competition.	From about the 4th year after planting, even in dry sites
mechanical weed control	Good work is possible only if there is little weed infestation (start early). Conserves soil moisture effectively, weeds are left near the trunk. Protect young fruit trees with pegs, if necessary, and keep the work as shallow as possible, so as to avoid root damage.	On lighter soils with few stones
covering with bark	Good suppression of weeds propagated by seed, effect lasts 3–4 years. Helps humus production and prevents drying out and extreme temperature fluctuations in the soil. Only put bark on well drained soil. Layer thickness 10 cm, width of strip 1.20 m.	Sites with light soils with low humus content, tending to be dry in summer
covering with rapeseed straw	Effect on weeds only lasts 1–2 years. Straw can be put down simply by hand or with a machine. Beneficial effects on soil similar to those of bark, but little increase in humus. Covering once provides about 100 kg of potassium/ha per annum. Layer thickness 15–25 cm, width of strip 1.20 m; 20 bales per 100 m.	Not on soils with excessive potassium
covering with water-permeable mulch film	100% effect against weeds, laying can be mechanized. Increases the moisture content of the soil in a similar way to hoeing. Wear-resistant water-permeable films available. Films present disposal problems, relatively high costs.	Only in orchards with good protection against mice, 1st–4th year after planting
wood chips, various compost materials	Unsuitable.	

Hydraulically driven **flat-shares**, which cut down vegetation just below the soil surface and shred it with an additional rotor. The ground can get churned up if the soil is wet. If it stays wet after use of the equipment the unwanted plants may start to grow again. The effect is unsatisfactory if the fruit trees are planted too close together.

Rotary cutters sometimes give very good results for weed control. The ground is finely broken up by revolving discs with angled blades or stars and is mixed with the weeds cut down so as to achieve better decomposition. These machines work satisfactorily even when the vegetation is high. They are expensive to buy, so it is advisable to get them through a machinery cooperative or joint investment scheme.

Disc harrows with sensors till the soil either in the direction away from the tree or in the direction towards the tree. This causes considerable displacement of the soil. The simple technology and favourable purchase price have led to increasing use of these machines recently. Disc harrows are more suitable for light soils.

Fertilizer application

Under the organic guidelines, commercial fertilizers and fertilizers produced by the grower can only be used on the basis of soil analyses and in accordance with the needs of the crops. Only then is it possible to calculate the amount still required and apply it in the form of permitted fertilizers. Well-balanced fertilizer application enables the soil to make nutrients available to the fruit trees in a harmonious ratio and in adequate amounts at the right time, and helps to keep the trees in physiological balance. If fruit producers want to apply fertilizers in accordance with requirements, they need to know the nutrient content of their soil before they start fertilizer application.

Nutrient requirements of pome fruit orchards

The nutrients that the plants take from the soil to grow and produce fruit need to be replaced by replenishing the soil and by the application of fertilizers. The nutrient uptake of an apple orchard in full production is relatively low. If the yield is between 25 and 50 t/ha, the trees take up 20–30 kg N, 5–15 kg P_2O_5, 50–80 kg K_2O, 17–20 kg CaO and 6–8 kg MgO (Greenham, 1980). The nutrients supplied by fertilizers are not fully available for use by the fruit plants, as nutrients may be lost as a result of leaching, runoff and fixation. On the other hand, nutrients are constantly being made available to the plants through mineralization and the action of weather on the soil.

Self-regulation of nutrient assimilation

In the soil close to the roots (**rhizosphere**) there is much greater release of potassium, phosphorus and nitrogen than in the soil that is a long way

Fig. 4.1. Perfekt inter-row sweeper.

Fig. 4.2. Ladurner clod breaker.

Fig. 4.3. Müller RPM inter-row cleaner.

Fig. 4.4. Humus Planet clod breaker.

Fig. 4.5. Clemens Radius 859 flat share.

Fig. 4.6. Spedo TPE disc harrow.

from the roots. Chemical and microbiological processes are equally important in this. In apple trees an infection of the root of the tree with the common soil fungus *Mycorrhiza* is of particular importance for phosphorus supply. However, the *Mycorrhiza* fungus is not the only means the apple tree has of adapting to reduced nutrient availability. As a general rule there are two other possibilities:

- **improved root growth**: increased root surface and thus improved spatial availability of nutrients
- **changes in pH** or root secretions in the rhizosphere.

Methods for determining nutrient requirements

Soil testing

Soil testing is a widely used means of finding out the specific properties and nutrient content of a soil and monitoring them over a period of time (see Chapter 2: *Planning and setting up an organic production unit*). A fertilizer programme can be drawn up on the basis of the results of soil analysis.

The availability of nutrients in the soil depends on the water regime, the depth of soil that can be used by the plants, the type of soil, the proportion of stones and the biological activity. Soil analysis, however, provides hardly any information about the physical and biological state of the soil. In spite of high nutrient levels, the availability of nutrients may be reduced by soil compaction and waterlogging. Evaluation of the soil profile or of spade samples is therefore of particular importance in deciding on the need for soil improvement measures (see section on *Soil preparation* in Chapter 2). Soil analyses can be divided into two groups. The first group allows a general impression of the soil to be obtained: **type of soil** (clay content), **humus content** and **pH**. The second group determines the nutrient status of the soil: the water-soluble nutrients that are available to the plants, **phosphorus** (P), **potassium** (K), **magnesium** (Mg) and **calcium** (Ca), and the reserves of P, K, Ca and Mg, by CAL, DL or AL extraction. The results of both analyses are used to calculate the fertilizers needed. In addition, the **biological activity** of the soil is determined in biospecific routine tests.

THE N_{MIN} METHOD FOR DETERMINING N REQUIREMENTS. Mineral nitrogens (N_{min}) include **nitrate** (NO_3^-), **nitrite** and **ammonia** (NH_4^+). Nitrite is an intermediate in nitrification and denitrification and occurs in negligible amounts in the soil. The amount of ammonia is highly dependent on soil pH and increases with increasing acidity.

TIMING OF N_{MIN} ANALYSES. N_{min} analyses should be performed around the end of March or beginning of April. The factor which limits yield is N_{min} availability in the spring. During the main growth phase from mid-May to mid-July the replenishment of N is not always sufficient to fully cover the needs of the trees. An adequate N_{min} supply should therefore be available for the apple blossom.

TAKING SAMPLES.   The soil specimen should be taken from two horizons (0–20 cm, 20–40 cm) in the planting strip. The main mass of the tree roots is at a depth of 0–40 cm. After the sample is taken it must be brought quickly to the testing laboratory. Refrigeration bags are convenient for transporting samples. The results of the analysis are interpreted in graph form.

For the commercial grower there is the possibility of a semi-quantitative analysis using nitrate test strips (produced by Merck) which show different intensities of colour depending on the nitrate content of the soil solution. The nitrate content is assessed by comparison with a reference colour scale. More precise determination of the nitrate content is possible using a photometer (e.g. Reflektoquant analyser) based on the reflectometry principle.

Determination of nutrient requirements by analysis of leaves

Apart from soil testing, an analysis of plants can also provide information on the nutritional status of fruit plants under certain conditions. It indicates the amounts of nutrient directly taken up and thus yields information on the actual nutritional status of the fruit plants tested. Leaf analyses should only supplement soil analyses, not replace them.

Table 4.2. Comparison of soil analysis and leaf analysis.

	Soil analysis	Leaf analysis
methods	different	uniform
extraction	nutrients available to plants	total nutrients
comparability	only partial	yes
effects of site on nutrient uptake	partially taken into account	taken into account
taking samples	laborious	simpler
analysis	longer than 1 week	shorter than 1 week
determination of nutrient requirements	for several years	snapshot

TIMING OF LEAF ANALYSES.   In practice, the following two periods have been found to be suitable for taking samples for leaf analysis.

1. **Early** leaf analysis (end of May to start of June).
2. **Late** leaf analysis (end of July to start of August).

Because the results of the analysis are obtained relatively late (in August), it is virtually impossible to take corrective measures for the same growing season. For this reason, late leaf analysis is now relatively little used in practice, and there is a clear preference for early leaf analysis.

WHAT ARE THE ADVANTAGES OF EARLY LEAF ANALYSIS?

- Leaf analysis at the time of the highest nutrient requirements gives a better indication of deficiencies or levels in excess of requirements.
- An imbalance in nutrient supply has greater effects in this growth phase than later.
- The fruit grower still has time to correct any nutrient deficiencies with fertilizers in the same growing season.

INTERPRETING THE RESULTS OF THE ANALYSIS

- The evaluation of the results of leaf analyses should not be governed by a rigid procedure
- various PC-based interpretation schemes are available
- the interpretation should be performed by or together with the extension officer
- graphic display of the results makes interpretation easier
- the needs of different cultivars should be borne in mind
- the following factors should be taken into account in the interpretation:
  - the weather in the last 2 weeks
  - shoot growth – severity of pruning
  - the fruit crop load
  - foliar fertilizer application
  - the results of the soil analysis
  - observations made among the plants (leaf development, leaf colour, etc.).

TAKING SAMPLES. A sample comprises 100 leaves – the two middle leaves of short or long shoots should always be taken. Leaves should be taken from two shoots per tree, giving a total of 25 trees per orchard. Average trees with normal fruit crop loads should be chosen. Shoots from the middle part of the crown of the tree should be selected; the leaves must get plenty of light. Samples should not be taken after heavy rain, intensive irrigation or application of foliar fertilizer. The samples should be packed in air-permeable bags, with an accompanying note, and taken or sent to the testing laboratory. Careful sampling is a key factor in determining how much information can be obtained from the leaf analysis.

Application of fertilizers in organic fruit production in practice

As a matter of principle, in organic farming the fullest use must be made of all measures available for improving soil fertility. These include careful preparation of the soil, avoidance of damage to the soil in tilling, optimization of humus levels, and application of farm-produced fertilizers. Optimum soil fertility ensures that the crop has an adequate supply of nutrients. Readily soluble N, P and Ca fertilizers are not allowed in organic fruit production; nor are K fertilizers containing chlorine. The commercial fertilizers suitable for organic use under *EC Regulation 2092/91, Annex II*, can be used only if there is a proven deficiency, but they cannot be used as replenishment fertilizers. The fertilizers and soil conditioners permitted for use in organic farming are listed in the Appendix.

Application of nitrogen fertilizers in organic fruit production

The annual nitrogen requirements of fruit trees are relatively low – only about 30 kg N/ha. An inadequate supply of N in the critical phase after flowering (cell division phase) can considerably reduce lignification, fruit set, and flower bud production and quality. Conversely, even a slight excess of N can lead to excessive shoot growth and the risk of physiological disorders, associated with keeping quality problems. The incidence of pests and diseases is also increased by excessive application of nitrogen fertilizer.

The N requirements of fruit trees vary within the growing season and are essentially limited to the months April to July, with a peak in June. Half of the N requirements in spring are met from the tree's own reserves. The provision of N to meet requirements is thus more a matter of timing than of quantity. Only a few N fertilizers, which are mineralized relatively quickly, are permitted in organic farming. This is why the application of N fertilizers needs to be precisely timed. Any competition from companion plants in the row of trees must be controlled in time, e.g. by mechanical tilling, so that mineralization is accelerated and the nitrogen is available for use by the trees. The N supply of an orchard should be optimized by means of N_{min} **analyses**.

Table 4.3. Guidelines for nitrogen fertilizer application based on N_{min}.

N_{min} value	Nitrogen replenishment	Recommended N fertilizer (kg/ha)
< 30 kg/ha	moderate	30–50 kg
	good	< 30 kg
30–50 kg/ha	moderate	0–30 kg
	good	0 kg
> 50 kg/ha	good	0 kg

GENERAL COMMENTS ON N_{MIN} ANALYSES.   There is a close relationship between N_{min} values and the humus content of the soil. The higher the humus content, the greater the amount of mineral nitrogen available to the plant. Mineralization of nitrogen generally increases from mid-April onwards. If the N_{min} content of the soil is low in the spring (<30 kg/ha), the levels of nitrogen in the leaves and fruit also tend to be low. Conversely, high N_{min} values in spring (>50 kg/ha) give rise to high nitrogen contents in the leaves and fruit. Depending on the humus content, 2000–10,000 kg of pure nitrogen/ha is organically bound in the soil. On average, mineralization releases 70–140 kg/ha of pure nitrogen (N) annually and makes it available to the crops. Nature thus provides us annually with about 300–500 kg/ha of a 27% nitrogen fertilizer, which must be taken into account when assessing total nitrogen (the rule of thumb is that **1% humus can provide about 30–40 kg of nitrogen/ha annually**).

Unfortunately the seasonal availability of nitrogen is often out of synchronization with requirements. This means that the supply of nitrogen is inadequate in certain growth periods, especially in the phase after flowering.

Lime and calcium fertilizers

For the production of quality fruit that keeps well, it is of enormous importance that the fruit trees have an adequate supply of calcium (Ca). Although only about 3 kg of Ca per hectare and crop is deposited in the flesh of the fruit, especially in the cell walls of the fruit, even slight Ca deficiencies lead to unstable cell walls and physiological disorders. Only slow-acting lime fertilizers, such as calcium carbonate (lime marl, marine algae), calcium–magnesium carbonate (dolomite) and calcium silicate, are permitted in organic farming. Lime fertilizers ($CaCO_3$) are the principal fertilizers used to supply Ca. They are used mainly to raise the pH of the soil and to improve and maintain its structure. Because of their corrosive effect on soil organisms, fast-acting lime fertilizers such as quicklime ($Ca(OH)_2$ or CaO) are not allowed. Calcium chloride ($CaCl_2$) can be used as a foliar fertilizer if there is demonstrable calcium deficiency.

Important organic fertilizers for fruit growing

Organic fertilizers for soil conditioning

Organic fertilizers produced *in situ* are primarily soil fertilizers and also – because of their organic matter content – precursors for humus production. If soil analysis reveals that humus levels are too low, application of organic fertilizer is highly advisable. Use of organic fertilizers produced *in situ* is the best approach in terms of the efficient recycling of resources. It is very important to take the use of organic fertilizers into account in the fertilizer application programme.

Humus fertilizers produced in situ

FARMYARD MANURE. Apply well-rotted farmyard manure in late autumn, winter; about 20–30 t/ha. Small fruit and all shallow-rooting rootstocks (M9, M27, quince C for pears) respond particularly well to this treatment. For young trees, covering the planting strip with farmyard manure has been found to give good results.

COMPOST

- Ideal source of humus
- humus built up by organic matter
- rich in nutrients and soil organisms
- increases the biological activity in the soil
- slow release of nutrients
- reduction of nutrient leaching
- additional material obtained from biogenic waste
- mineral fertilizers used less or not used at all.

APPLICATION OVER THE WHOLE AREA OR ON THE PLANTING STRIP.   After ground-levelling work which exposes a lot of dead soil, this supplies humus and quickly brings the soil back to life. After earth-moving work it is advisable to 'inoculate' the whole area with farmyard manure or compost ($30–40$ m^3/ha, $2–3$ kg/m^2) and thus ensure that the soil is brought back to life. Because of their variable composition, composts should be analysed before use and the nutrient content taken into account in the comprehensive record of fertilizer application.

ADDITION OF FERTILIZERS IN THE PLANTING HOLE.   When planting, use $10–20$ litres per plant.

Organic fertilizers for supplying nutrients

Apart from organic fertilizers produced *in situ*, the main products available are those derived from animal waste. Recently, because of a shortage of animal waste, use has also been made of plant residues, especially from fruits containing oil – notably oilseed residues (rapeseed, sunflower, castor beans, grape pips, olives, groundnuts, etc.). Marine algae are also becoming increasingly important because of their high proportion of trace elements and their amino acid content.

Castor oilmeal

This is the residue left after pressing the seeds of the castor oil plant. Castor oilmeal is mainly used as a nitrogen fertilizer and soil activator. The amounts used depend on the results of the N_{min} analyses and the needs of the crops, and range from 800 to 1500 kg/ha.

Rapeseed, groundnut, sunflower oilmeal

These oilmeals all still have a high oil content even after pressing; the bulk fibre component is subsequently converted to humus. All these fertilizers are considered to be rich in nitrogen.

Table 4.4. Protein and total nitrogen contents of various oil-containing fertilizers.

Fertilizer	Protein (%)	Total N (%)
rapeseed cake	35	5.2
linseed cake	32–36	4.7–5.3
thistle cake	32–35	4.7–5.2
sunflower cake	35–38	5.2–5.6
hemp cake	30–40	4.4–5.9
poppy seed cake	30–40	4.4–5.9
pumpkin seed cake	57–60	8.4–8.8
sesame cake	35–40	5.2–5.9
groundnut cake	60–65	8.8–9.6
walnut, hazelnut cake	50–55	7.4–8.1

Limes, rock meals and clay

These substances are also used as organic fertilizers. They are intended mainly for soil conditioning. They can also be used to improve composts in a natural way.

Fertilizers derived from animal waste

MEAT MEALS AND BLOOD MEALS. Meat meals and blood meals are available in finely powdered form and are both fairly fast-acting high-quality organic nitrogen fertilizers (9 and 12% N, respectively). There are problems with dust development during application. These fertilizers should only be applied when there is no wind. In addition, during application of the fertilizers it is advisable to wear dust masks or use tractors with climatized cabins. Any fertilizer spreaders which can precisely distribute fertilizers in powder form are suitable for applying them.

HORN MEALS. These are obtained from ground animal horns and hooves and are also fairly fast-acting, with the fineness of the grinding being important for the relevant application. The most finely ground meals are very fast-acting (within 10–14 days, but drifting in the wind can be a problem), while the meals with medium particle size take effect in about 4 weeks and the coarsest meals in about 8–10 weeks. If all three fractions are mixed together, a source of nitrogen which lasts for the entire season is obtained.

BONE MEALS. These are particularly suitable for supplying phosphorus and lime to the soil. Bone meals are produced from bone waste from abattoirs. Before grinding they are treated with hot steam and the fat is removed. The fat should not be removed with tetrachloromethane or similar toxic substances, however. If it is, the bone meal is not suitable for use on organic crops.

Table 4.5. Organic fertilizers (composition as percentage of dry matter).

Fertilizer	Organic matter	N	P$_2$O$_5$	K$_2$O	MgO
horn meal	80–90	12–14	5	0.2–0.8	0.6–1.2
castor oilmeal	75	5–6	2	1	0.3
rapeseed cake	70	5	0.9	1.5	–
vinasse	50	3.5	0.4	7.5	–
meat meal	90	9	5	0	–
low-potassium Biosol	85	6.5	1	1.5	–
blood meal	85	12	0.2–2.5	0.4–1.7	0.2–0.5
hair or feather meal	75–80	13	–	–	–
bone meal:					
with fat removed	45	6	15	0.1	0.6
with glue removed	15	1	31	0.3	0.6
straw	60	0.4	0.2	1	0.1
composted bark	50	0.3	0.1	0.2	0.1
farmyard manure	20–25	0.5	0.3	0.5	0.2
brewers' grains	60	0.45	0.35	0.13	–
prunings	60	0.8	0.25	0.8	0.2
green prunings compost	25	0.2	0.1	0.15	0.4
guano	50	6–7	4–6	2.0–3.0	3–4.5
dried poultry manure	60	4	5	3	

Foliar fertilizers in organic fruit growing

Animal proteins in liquid formulations

Amino foliar fertilizers contain 55% amino acids and peptides and 9% organically bound nitrogen, as well as the nutrients and trace elements naturally contained in animal protein. Short-term deficiencies can be made good by using these foliar fertilizers at the stage when N requirements are highest (end of flowering, T stage). It is advisable to apply them only when leaf or soil analyses reveal nitrogen deficiency. These products cannot be mixed with products containing mineral oil or with copper. They also have the advantage that they improve the adhesion of plant protection products and foliar fertilizers.

Table 4.6. Examples of products and applications.

Product/manufacturer	Application recommended by manufacturer
Aminosol (Lebosol Dünger)	Pome, stone and small fruit: 0.5–1.0% (second flowering of pome fruit only 0.2%)
Siapton 10 L (Siapa)	Application: 0.3–1.0%

Vinasse liquid fertilizer

Vinasse is an organic liquid fertilizer which is obtained as a byproduct in sugarbeet processing. It contains 3–5% nitrogen, 0.14–2.0% phosphate and 5.0–7.5% potassium, together with trace elements, vitamins and enzymes. Care should be taken when mixing with sulphur and copper in high concentrations.

Table 4.7. Examples of products and applications.

Product/manufacturer	Application recommended by manufacturer
BioTrissol (Neudorff)	Application: once or twice at first flowering, 2–3 times at second flowering, 5 l/ha
Provita–natural vinasse	Application: 0.3–0.5%

The following foliar fertilizers can also be used if there is demonstrable deficiency:

- **calcium chloride** in cases of calcium deficiency, against bitter pit and physiological disorders, etc.
- **magnesium sulphate** in cases of magnesium deficiency
- **trace elements**: boron, zinc, iron, copper, manganese and molybdenum in the form of sulphates and chelates.

Thinning in organic fruit growing

Thinning means the removal of excess blossoms or fruits. The aims of thinning are firstly to **prevent biennial bearing** and secondly to **improve internal and external fruit quality**.

Prevention of biennial bearing

One of the most important aims of thinning is to prevent biennial bearing, i.e. to maintain regular fertility.

In favourable years (when there is a surplus of assimilates and a light fruit crop) a tree produces a large number of flower buds. In the next year (**bearing year**) it then suffers from a shortage of assimilates, as they have been completely used up by the young fruit. At the same time the young fruitlets produce phytohormones (auxins, gibberellins) which also inhibit the production of flower buds. For this reason the tree does not produce any flower buds, or only a few. The consequence is a **rest year** in which the tree bears few fruit. This gives rise to a surplus of assimilates which once again causes the tree to produce too many flower buds.

Biennial bearing can also be caused by a late frost, which destroys all the flower buds in spring. Biennial bearing is a common problem in apples, pears and plums, but is virtually unknown in cherries.

This condition of biennial bearing frequently persists for many years unless man intervenes by removing excess blossoms or fruits. As a result the tree still has sufficient assimilates to produce flower buds for the next year, and there is no inhibition of flower buds by phytohormones which are synthesized in the seed of the young fruitlets. If thinning is successful in inducing biennially cropping cultivars to crop regularly, average **yields can be raised by 15–20%** with a simultaneous improvement in quality. Trees are usually not thinned enough to correct biennial bearing. If there is a good blossom, 5–10% of the blossoms which develop into fruit are sufficient for a full yield. The critical factor for regular cropping of fruit trees is the ratio of the leaves to the number of fruit. **The optimum leaves/fruit ratio for preventing biennial bearing is about 30–40:1** (i.e. about 30–40 leaves per fruit). A bigger leaf mass is needed to induce regular cropping in cultivars with pronounced biennial bearing. In a large number of trials, complete thinning of half one side of the tree proved to be a very effective measure for breaking the biennial bearing cycle.

Fig. 4.7. Biennial bearing is one of the biggest problems in organic fruit growing.

Enhancing fruit quality

The external quality, and thus the price, of fruit depends to a great extent on the size of the fruit. A producer who wants to achieve good returns for his fruit has to produce consignments with fruit of good quality and a high proportion of large-sized fruit. Normally fruit commands a higher price on the market as quality and size increases. Fruit sizes which are typical for the cultivar can only be achieved with an optimum fruit crop load, however.

Since the nutrition of the fruit largely depends on the assimilates available, a minimum number of leaves must be present for each fruit in order to ensure optimum fruit quality. In the case of apples, for example, about **20–30 fully developed leaves per fruit are needed for good fruit quality**, and these leaves must be present on the relevant fruit shoot. The leaves on long shoots do not count towards this total. A greater number of leaves is generally needed in the case of cultivars with large fruit and small leaves, and in parts of the crown that receive little light. If there are fewer fruit on the tree, this will have a positive effect on the leaves/fruit ratio. Excess fruits must be removed, otherwise fruit size will be severely affected. Quality is determined not only by fruit size, however, but also by the morphological and biochemical development of a fruit. Poorly developed fruits are removed when thinning. Improvements in quality are primarily achieved by manual thinning. Large numbers of trials have clearly demonstrated the economic importance of thinning for raising quality. The more fruits a tree has to provide for, the poorer is the internal quality of these fruits (in terms of both composition and flavour).

Timing of thinning

The time when thinning is performed can vary widely, depending on the intended purpose. After the June drop, thinning has very little effect on biennial bearing. The best approach is to remove the blossoms (blossom thinning). In this way it is possible to conserve many of the assimilates which would otherwise have been needed for the development of the young fruitlets. Early thinning does involve some risk, however, because it is impossible to predict the extent of any subsequent damage from late frosts and the extent of the natural fruit drop. Unfortunately this undoubtedly significant risk is unavoidable if biennial bearing is to be corrected.

If thinning is only intended to improve fruit quality, it can also be performed after the June drop. A significant improvement in quality occurs even when thinning is done a few weeks before harvesting. In the case of apples and pears, thinning still has a beneficial effect up to about 4 weeks before harvesting; in early cultivars, picking the first ripe fruits 2 weeks before the main harvest can significantly improve the quality of the remainder of the crop.

Thinning methods

The number of blossoms can already be reduced when pruning the tree, by heavier pruning of the older spurs. So as not to stimulate growth unnecessarily, however, the tree should not be pruned too severely. There are several different ways of performing the thinning operation in organic fruit production:

- manual thinning
- mechanical thinning
- chemical thinning with organic thinning agents (organic fungicides and insecticides).

Hand thinning

BLOSSOM THINNING.   The most reliable method of inducing regular crop-ping in biennially cropping cultivars is **blossom thinning by hand**. It is advisable to make an assessment of flowering intensity (giving a score from 1 to 9), however, before the blossom clusters are removed (with secateurs or by pinching out with the fingers). Only trees with a **flower-ing intensity score of 7–9** should be blossom-thinned. The best time for this very effective measure is the red bud to balloon stage. To break the biennial bearing cycle, at least half to three-quarters of the blossom clus-ters should be removed. This measure is only advisable for smaller trees, as from a certain tree size too much manual labour is required. Trees with a **flowering intensity score of 5** do not need thinning and exhibit the same flowering intensity in the following year (trees with stable flow-ering in physiological equilibrium).

Table 4.8. Flowering intensity scoring system for deciding on the degree of thinning.

Flowering intensity	Number of blossoms	Thinning
1	none	none
2	very few	none
3	few	none
4	few to medium	none
5	medium	if necessary
6	medium to high	yes
7	high	yes
8	very high	yes
9	completely covering the tree	yes

FRUIT THINNING.   Hand thinning after the June drop serves mainly to improve the quality (external and internal quality) of the fruit. Thinning to break the biennial bearing cycle (mechanical, chemical or manual) around or shortly after flowering cannot replace hand thinning after the June drop. The following positive effects are achieved by hand thinning:

- better fruit size
- higher proportion of surface colour
- higher level of constituents such as sugar, acid and vitamins (more flavour)
- firmer fruit

- increasing propensity of the trees to flower
- bigger crop
- reduced sorting costs.

Table 4.9. Advantages and disadvantages of hand thinning.

Advantages of hand thinning	Disadvantages of hand thinning
low risk	very labour-intensive
poorly developed fruits can be removed	no effect on biennial bearing if done late
time can be chosen as required	

Recently, organic fruit growers have started to combine mechanical and hand thinning when there is heavy flowering:

- Excess buds or blossom clusters are removed before flowering **using the thinning machine**. In this way the strain on the trees is reduced at an early stage (breaking the biennial bearing cycle).
- In order to achieve high fruit quality, **additional thinning by hand** is essential immediately after the June drop.
- In August, a further picking of underdeveloped fruits is beneficial.

Because of the high labour requirements, working out at 100–300 hours/ha, hand thinning alone is not usually a feasible proposition. It is the only suitable thinning method, however, for young orchards, for cultivars susceptible to bitter pit with not very heavy flowering, and for additional thinning of heavy-cropping cultivars where chemical and/or mechanical thinning has been unsatisfactory.

The following points need to be borne in mind in the hand thinning of pome fruit:

- The fruit stems should be left on the tree.
- Frequently only one fruit is left per blossom cluster. It is better, however, to remove some blossom or fruit clusters completely (because the tree is then likely to produce flower buds at that point) and to leave **2 or 3 fruits on the remaining clusters**. In this way the shoots from which all clusters have been removed have the assimilates available for producing flower buds. Nor is there any inhibition of flower bud production by phytohormones (which are synthesized in the seeds of the young fruitlets). Biennial bearing can thus be reduced by hand thinning an at early stage.
- The centre blossom often gives to **poorly shaped fruits**, which should be removed when thinning.
- The fruits which are **smallest in size** and lagging behind in development (as can be seen from the pointed calyx part) should also be removed in the thinning operation.
- **Poorly developed fruits** (especially from the inner part of the tree) should be removed during hand thinning. In particular these are:

- fruits which are **damaged, russeted** and affected by **disease** (scab, mildew) or pests (with areas of insect damage, wind rub, hail damage, etc.)
- poorly fertilized, asymmetrical fruits
- fruits which have received little sun
- fruits on branches that hang down too much.

The consequence of poor exposure to sunlight and inadequate sap supply is underdeveloped fruits. Although these can ripen at the same time as fully developed fruits, they are of inferior quality. Their shape is usually not the typical shape for the cultivar. Harvesting after the June drop is possible with careful hand thinning. **Large fruits should be left on the tree**, even if there are 2–3 in one cluster. In cultivars that are susceptible to bitter pit, such as Braeburn, Gloster, Boskoop and Jonagold, care should be taken to make sure that fruit thinning is not too severe. As a result only fruits that are too big would develop, with increased susceptibility to **physiological disorders** (and reduced shelf life). In these cultivars there should not be any single fruits on vigorous shoots which receive plenty of sunlight and can sustain more than one fruit, as single fruits get too big and are of inferior quality.

HOW SHOULD HAND THINNING BE CARRIED OUT?   The table below shows how much a tree must bear in order to achieve the desired yields at different planting densities. Crop yields from 20 to 50 t/ha were assumed. The individual grower can substitute his own figures if he expects higher yields.

Table 4.10. Individual tree yields as a function of area per tree and crop yield.

Tree spacing	Area/tree (m²)	Trees/ha (9000 m²)	Crop yield (kg/ha)			
			20,000	30,000	40,000	50,000
4.0 m × 2.0 m	8.00	1125	17.8	26.7	35.6	44.4
4.0 m × 1.5 m	6.00	1500	13.3	20.0	26.7	33.3
3.5 m × 1.25 m	4.38	2057	9.7	14.6	19.4	24.3
3.0 m × 1.0 m	3.00	3000	6.7	10.0	13.3	16.7
3.0 m × 0.75 m	2.24	4000	5.0	7.5	10.0	12.5

Table 4.11. Relationships between fruit diameter, fruit weight and number of fruits/kg in different cultivars.

Fruit diameter (mm)	Fruit weight (g)	Number of fruits per kg	Cultivars
60	92	11	
65	120	8	
70	145	7	Rubinette
75	180	6	Gala, Arlet, Elstar
80	215	5	Golden Delicious, Gloster, Idared
85	260	4	Boskoop, Jonagold

Below is an **example for Golden Delicious** with 3000 trees/ha:

Tree spacing:	3 m × 1 m
Expected crop yield:	30,000 kg/ha
Individual tree yield:	10 kg/tree
Fruit size:	75–80 mm = 5–6 fruits per kilogram
Fruits per tree:	55

This means that 55 fruits are needed for a fruit size of 75–80 mm, in order to harvest 30 t/ha at a tree spacing of 3 m × 1 m. This is a model calculation of course; the corresponding data must be substituted for other tree spacings, cultivars, fruit sizes and crop yields. In hand thinning, therefore, what is important is to **count the fruit**. The fruit grower must be able to estimate how many fruit there should be per tree.

It must also be borne in mind that as the number of trees per hectare decreases, each individual tree must bear a significantly higher number of fruit in order to attain the desired yields. Unavoidably, however, this can only be achieved at the expense of fruit quality, especially the internal quality (sugar and acid content and firmness of the flesh). With fewer fruit on the tree, the leaves/fruits ratio improves, and so quality is raised. The **specific crop load density** (number of fruit per cm^2 of cross-sectional area of trunk) is the determinative factor for fruit quality.

Fig. 4.8. Manual fruit thinning (before).

Fig. 4.9. Manual fruit thinning (after).

Mechanical thinning (thinning machine)

An interesting alternative to manual or chemical fruit thinning could be mechanical blossom thinning using a tractor-mountable thinning machine. For the following reasons there is an urgent need for new thinning techniques:

- no chemical thinning agents have yet been approved in organic fruit growing
- inadequate effect of biological thinning agents
- early thinning has the greatest effect in breaking the biennial bearing cycle.

MODE OF OPERATION.   The machine consists of a carrier on which a verti-cally rotating spindle is mounted. On this spindle there are synthetic fibres which can easily be fitted and removed again. By virtue of its adjustable angle of inclination, the spindle can be adapted to different tree forms. It is hydraulically driven and moved along with the tractor on both sides of the tree. Through the rotary movement of the spindle, the synthetic fibres knock off individual blossoms or blossom clusters, especially at the periphery of the tree. There is the possibility of rear or front mounting. The degree of thinning is determined by the speed of travel, the rotation rate of the spindle and the number of synthetic fibres used. The thinning machine gives uniform and good thinning with small tree forms (super-spindle and crown forms with no excessively strong framework branches in the tramline) and trees with little shoot growth (with a lot of whorled wood). Current experience suggests that the opti-mum speed of travel is 6–8 km/h and the optimum spindle rotation rate is 300 rev/min.

TIME WHEN THE THINNING MACHINE SHOULD BE USED.   Green bud–red bud stage; later treatments lead to damage to fruit, while earlier treatments lead to increased damage to the wood and often do not have sufficient effect.

POSSIBLE DISADVANTAGES.

- Can only be used for certain tree forms (slender spindle without strong lateral branches)
- damage to wood
- early use (a problem if there is subsequent frost damage to blossoms)
- encourages and spreads diseases and pests (fireblight, canker, woolly aphid)
- effects on shoot growth.

Chemical blossom and fruit thinning

No registered chemical thinning agents are available in organic fruit growing for blossom or fruit thinning. Certain organic fungicides and insecticides, however, have some thinning effect by inducing increased fruit drop as a result of their corrosive action.

NEUDOSAN.   Neudosan is a product based on potassium salts of natural fatty acids. In accordance with the manufacturer's recommendations it was used at 2–4% at full bloom with 1000 litres of water/ha. Although a good

Fig. 4.10. Relation between stem growth and fruit set.

Fig. 4.11. Mechanical thinning with a thinning machine.

thinning effect was observed, the success achieved was not satisfactory, as there was excessive russeting of fruit in all the cultivars tested.

LIME SULPHUR SOLUTION. Lime sulphur solution is one of the most important fungicides in organic fruit growing. A thinning effect was observed in some trials. Use of 20–40 kg/ha (2.0–4.0%) at full bloom is recommended. Cultivar sensitivity and possible side effects on beneficial fauna should be determined by careful experimentation.

Growth-regulating measures in organic fruit growing

All growth-regulating measures in organic fruit growing must have the aim of improving the **penetration of light** to the leaves and the fruit (weak rootstocks, small tree forms, not too dense planting, pruning, etc.). A loose crown structure (long pruning), obtaining only moderately vigorous production of new shoots, and moderate pruning in summer contribute to better fruit quality and reduce the risk of disease and pests.

Pruning

The primary aim of all pruning in fruit growing is to create and maintain an equilibrium between shoot growth and yield **(physiological equilibrium)** as a basis for early, high, regular and high-quality yields. There is physiological equilibrium if, in addition to a heavy fruit crop, a tree also

Fig. 4.12. Trees with too strong shoot growth are not in physiological equilibrium and will not give the best possible yield.

shows a necessary minimum growth. Only trees in this state of equilibrium regularly give fruit of the highest external and internal quality (size, shape, colour and composition) with good storage life.

Winter pruning

The pruning of fruit trees, like all other cultural measures, must be adapted to the cultivar, site and physiological condition of the trees. The **optimum shoot length** for production of a large number of high-quality flower buds depends on the fruit species and in apples ranges **from 5 cm to a maximum of 30 cm** ('quiet tree'). Many apple orchards often have problems with too vigorous shoot growth, combined with unsatisfactory yields and increased occurrence of biennial bearing. Vigorous shoot growth is often synonymous with poor exposure to light, unsatisfactory and irregular yields, poor fruit colour and high susceptibility of the fruit to physiological disorders (e.g. bitter pit, internal browning). **Increased incidence of disease and pests** (e.g. scab, mildew, aphids) can also be seen in orchards of this type. Trees on vigorous rootstocks (M26 and above), which are still prevalent in organic fruit growing, need a different type of pruning from trees on M9 rootstocks. In orchards with rows under ground cover, growth on the M9 rootstock may be too weak because of the competition from the ground cover vegetation, so more drastic pruning for fruiting is required. In addition, pruning to control biennial bearing is needed more often in organic orchards.

Tearing of shoots

Tearing of vigorous young shoots before the lignification of the base of the shoot (end of May to mid-June) can be recommended (so there is no actual summer pruning). Summer shoots in the interior of the crown of the tree (vigorous shoots on arched branches, water shoots) can be removed quickly and simply by this measure. Apart from the tearing of summer shoots there is also the possibility of tearing lignified 1-year-old shoots or branches which are vigorous (stronger than half the diameter of the attachment point) and steeply angled, especially in the upper crown region. Unlike cutting, the tearing of shoots is a natural process and the wounds made by tearing generally heal better than wounds made by cutting.

Summer pruning

Summer pruning can be commenced as soon as the tree has finished shoot growth. The tree **should not produce further shoots** at the site of the cut, as these new shoots cannot be used to create the spindle. They are poorly lignified and thus susceptible to frost damage, mildew and scab. Regrowth shoots which carry scab spores (conidia) are a dangerous potential source of apple scab infection in the spring. **The most favourable period for summer pruning, therefore, is mid-August.** Tying and 'maypoling' is also best carried out at this time, but can be left until later, depending on weather conditions (rainfall and temperature), altitude, shoot growth and cultivar. The earlier that pruning is carried out, the more likely it is that the tree will respond by producing shoots in the same year. If summer pruning is done too early there may be premature bursting of the flower buds. Blossoms in September, however, are useless for the fruit grower and reduce the fruit set in the next year. For these reasons **early summer pruning in June or July is not recommended**.

Aims of summer pruning of fruit trees

TO CURB GROWTH.   Trees which grow vigorously (e.g. in a year when little or no fruit is produced) cannot have their growth curbed by drastic summer pruning. One possibility of curbing shoot growth is through **self-regulation of the tree by August pruning**. An orchard which is intended for August pruning should be left completely without pruning for 18 months (no summer pruning and no pruning in the following winter).

Fig. 4.13. Tearing of shoots.

Fig. 4.14. Summer pruning.

Fig. 4.15. Regrowth shoots as a result of summer pruning carried out too early.

- **Slight vegetative growth** (lots of possibilities for shoot production, so shoots are shorter) with early cessation of shoot growth, can already be seen after the first omission of winter pruning.
- Leaf size decreases relative to previous years, so the crowns **allow penetration of light and air** in spite of the extensive ramification.

After mid-August (cessation of shoot growth) the necessary corrections must be made to the trees, i.e. heavy pruning should now be carried out to compensate for the deliberate omission of pruning the previous winter. This sometimes means that whole branches and parts of branches bearing apples at an advanced stage of development are removed. The grower may be reluctant to do this and may have to overcome a certain psychological barrier. Weak, drooping branches with underdeveloped fruit should also be removed at this time.

ADJUSTMENT OF THE LEAVES/FRUIT RATIO. If the fruit crop is small and there is vigorous shoot growth (too many leaves per fruit; more than 50) the fruits become too big and thus susceptible to physiological disorders (e.g. bitter pit, glassiness). Fruits from trees which have been given a moderate summer pruning contain more calcium and less potassium, with a somewhat lower incidence of bitter pit as a result. The **ideal leaves/fruit ratio** for **good fruit quality** and **optimum storage life** of the fruit is about **40:1** (see section on *Fruit thinning*). If there is heavy fruit set, the number of fruit should be reduced by hand thinning.

MORE COLOUR. The increasing popularity of cultivars with a red colour (e.g. Jonagold, Elstar, Gala and Arlet) makes summer pruning even more beneficial. These cultivars develop additional colour if they get **more light just before harvesting** (pre-harvest pruning about 10 days before harvesting). Insolation of the fruit can easily be ensured by removing shoots that have a shading effect. This pruning should not be too early or too heavy (danger of sunscorch). Fruits in shade often fail to develop colour and are also susceptible to various physiological disorders (e.g. internal browning) which reduce shelf life. Summer pruning can help to improve internal and external fruit quality.

STRENGTHENING OF FLOWER BUDS. The improved penetration of light into the crown of the tree will lead to stronger flower buds and better chances of fruit set near the central leader.

POST-HARVEST PRUNING. In **autumn cultivars**, e.g. Elstar, Gala and Arlet, **in order to restrain growth**, part of the pruning (such as removal of entire branches) can be carried out after harvesting. Caution should be exercised, however, as there is a tendency to prune too heavily when the tree is in leaf.

Regulation of growth by careful choice of rootstock

(see Chapter 3 on *Apple, pear, plum and cherry rootstocks.*)

Regulation of growth by choosing the most suitable level of the graft union

The effect of the level of the graft union on growth is undisputed. This property of dwarfing rootstocks is used to obtain trees with less shoot

Table 4.12. Cox's Orange Pippin (susceptible to biennial bearing), planted winter 1992/93. Results in 1994–1996.

Rootstock	kg per tree 1994–1996	Weight of fruit (g)	Trunk circumference (cm) March 1997
T337-15[a]	21.8	184	11.3
T337-25	27.2	183	11.7
T337-35	26.4	184	11.9

[a] Height of graft union.

growth and greater yield (generative trees). The higher the level of the graft union on a dwarfing rootstock, the weaker is the shoot growth induced.

Regulation of growth by double-working with an interstock

Interstocks affect the growth of a fruit tree and thus also the yield. In apples, for example, the cultivar 'Zoete Aagt' is used as the interstock for Cox's O.P. and the cultivar 'Summer Red' is used as the interstock for Elstar. A further advantage of double-working is better ramification of the planting stock at an optimum height. The length of the interstock also has an influence on shoot growth. The length is usually between 30 and 35 cm.

Regulation of growth by ground cover management

(see section on *Soil care* in Chapter 2.)

Regulation of growth by root pruning

Root pruning is one of many ways of curbing growth, but it is not a cure-all. If growth has been curbed after root pruning, the fruit trees should be kept in physiological equilibrium by appropriate pruning of the tree (long pruning) and other cultural measures.

In orchards where a hail net is used the natural upwards development of the crown of the tree is often restricted by the net, with the result that in many cases the central leader has to be diverted at an early stage, before the top of the tree has slowed down in growth. In the vast majority of cases – without accompanying measures to curb growth – this results in increased vegetative growth. Root pruning in such cases is a quick and cost-effective growth regulation technique which can also be used in organic production. Shoot growth, and thus also fruit size, is reduced by cutting away part of the root with a root pruner. The effect of root pruning may vary from year to year.

Before starting root pruning, it is very important to estimate the distribution of roots in the soil (dig first, then prune). To prevent the risk of damage due to a sudden effect on growth and an excessive reduction in

Fig. 4.16. Double-working with Summer Red for growth regulation.

Fig. 4.17. Root pruner.

fruit size, root pruning should be carried out as a special measure with differentiation according to site (soil, climate and possibility of irrigation) and cultivar (cultivars with big or small fruit).

Possible pruning times

- March–April, if few flower buds are produced (biennial bearing), in cultivars with big fruit.
- Mid-June (after the T stage), if flower bud production is good, in cultivars with medium-sized or small fruit, or if there is no possibility of irrigation.

Extent and depth of pruning

Single row (on one or both sides): 30–50 cm (distance from the row) × 25–40 cm (depth of pruning).

Cutting into the trunk with a power saw

If there are major growth problems, making a wedge-shaped incision in the trunk with a power saw has been found to be effective as a method of curbing growth. The best time for doing this is between budburst and flowering. This method is used mainly on vigorous rootstock–cultivar combinations in apples and pears, e.g. on seedlings.

5 Plant protection

Principles and aims of organic plant protection

To understand the principles and rules of organic plant protection and the differences between the various plant protection systems it is first necessary to look at some principles of general ecology.

Why is plant protection needed at all?

In every natural environment (**natural ecosystem**) every organism has a particular function, as each individual organism is the link in a long chain (**food chain**). The ecosystem can survive in the long term only if this function is fulfilled. These interlinked functions are made possible by the biological equilibrium which exists between the individual organisms (**biocoenosis** = living community) and which as a result of many interactions determines the dynamics of the various populations. An organism is therefore simultaneously a prey and a predator and plays the ecological role of both the controller and the thing controlled.

On a piece of land that is farmed (**agricultural ecosystem**), on the other hand, the living community is highly simplified and often consists of only one cultivar which is being grown, and which because of its presence in very large quantities provides favourable conditions for its predators: **phytophagous** (plant-eating) and **phytopathogenic** (disease-causing) organisms. These can spread unchecked because they have no natural enemies, the biotope of the latter having been greatly reduced by agricultural specialization.

This means that whereas in the natural ecosystem all organisms are 'useful', in the agricultural ecosystem some organisms have a 'disturbing' effect. In the agricultural ecosystem the very same organisms that contribute to biological equilibrium in the natural ecosystem may be an unbalancing element which man has to eliminate or at least control by protective measures.

© CAB International 2003. *Organic Fruit Growing* (K. Lind, G. Lafer,
K. Schloffer, G. Innerhofer and H. Meister)

Table 5.1. Biocoenosis.

	Biocoenosis	
	Natural ecosystem	**Agricultural ecosystem**
made up of	wild plants	cultivated plants
	wild animals	farm animals
	decomposition organisms	human beings
characterized by	many species with few individuals	few species with many individuals
	wide biodiversity	little biodiversity
	complex living community	simple living community

Components of the agricultural ecosystem

The planted tree is the focal point of many factors that play a decisive role in determining the condition of the fruit crop and the quality and quantity of the product. Pest control is only one of the many factors involved. In the overall context of integrated fruit production, and especially in the context of organic fruit production, pest control has to be managed in such a way as to respect and enhance the positive factors in the agricultural ecosystem.

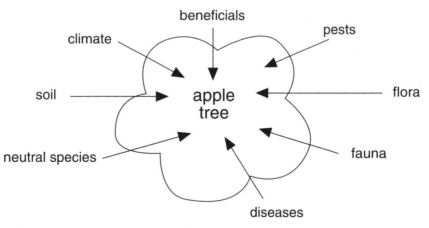

Fig. 5.1. Environmental factors affecting an apple tree.

Human intervention by various methods of plant protection

In organic fruit production, greater importance is attached to preventive methods of plant protection. Since the choice of pesticides is very restricted in organic production and they are often less effective than in integrated production, the best possible use must be made of all methods of plant protection in order to achieve efficient results.

Table 5.2. Methods of plant protection.

Cultural	Biological	Biotechnology	Physical	Chemical
choice of site	introduction of extraneous predators and parasites into the agricultural ecosystem	use of hormones	use of optical stimuli	use of chemical plant protection agents
choice of cultivar		use of pheromones (messenger substances secreted by insects)	use of acoustic stimuli	use of biological plant protection agents
fertilizers	protecting and improving the performance of existing parasites and		thermal radiation	
pruning	predators	use of deterrents and substances that induce feeding	mechanical agents	
	microbiological control			

Biological pest control takes advantage of the countless antagonistic relationships found in a natural environment in order to control the population of harmful organisms.

This is achieved in two ways:

- By control of the environment, with the aim of protecting and enhancing the effect of the existing natural antagonists.
- By propagation and introduction of natural antagonists.

The return to nature involves not only taking advantage of the complex defence capacity of plants (resistant varieties) but also applying practices that bring back a certain species diversity (crop rotation, mixed cropping, ground cover, hedges), thus restoring the lost antagonistic relationships and encouraging nature's own self-regulation mechanisms.

The few pesticides that are allowed in organic fruit growing are all of plant or mineral origin.

Encouraging biodiversity in orchards

In organic fruit growing, the main focus of attention should not be the individual treatments against various pests and diseases but the establishment of a diverse and largely self-regulating fauna and flora.

The following measures can be taken in the orchard in order to achieve biodiversity of flora and fauna:

- **Encouraging birds:**
  Hang up nesting boxes for blue-tits and coal-tits.
  Set up perches for birds of prey.

- **Encouraging predators and parasites:**
  Create wild-flower strips at the edge of the orchard.
  Alternate mulching and mowing of tramlines.

Many other possibilities, such as weed-infested rows, extensive meadows on unused areas or wild-flower strips in the tramlines, have to be rejected either because of increased vole risk or because of labour cost considerations.

Monitoring in the orchard

In order to be able to effect plant protection in a way that is as economical as possible and does the least possible harm to beneficials, it is important to precisely monitor the occurrence of pests and beneficials in the orchard. Direct control measures should be carried out only after the damage threshold is exceeded.

The following methods are available for determination of species composition and damage thresholds:

- **Examination of samples taken from branches in winter:**
  In winter older wood is examined mainly to see if there are any over-wintering stages of the San José scale or red spider mite.
  It is not necessary to count individual eggs.
  Zero tolerance is the rule with regard to San José scale.
  In the case of the red spider mite, treatment is necessary if the fingers are stained red by the winter eggs when wiping over the base of the branch.
  Winter eggs of the red spider mite are unlikely to be found, however, in an orchard with a stable balance between the red spider mite and predatory mites.

- **Knockdown test:**
  The animals present are knocked down from the branches with a knockdown funnel. This type of test is especially important in the case of the blossom weevil and should be carried out around mid-day, when the temperature is high (>12°C).
  This gives an idea of the total fauna of an orchard.

- **Visual inspection:**
  In this method 100 plant organs (blossom clusters, rosette leaves, long shoots, fruits, etc.) are selected at random and inspected with the aid of a good magnifying glass (10–15× magnification).

Checks are made to determine the level of infection or infestation and develop a control strategy.

Table 5.3. Important monitoring times.

Time	Infection/infestation checks	Efficacy checks
before flowering redbud–balloon stage	leafrollers, winter moths, noctuid moths, budrollers, aphids, sawfly, red spider mite	red spider mite
after flowering petal drop – June fruit drop	leafrollers, red spider mite, rust mites, aphids, apple sawfly, wood borers, scab	leafrollers, red spider mite, aphids, blossom weevil, sawfly
July	red spider mite, rust mites, scab	codling moths, fruit leafrollers
harvest	leafrollers, smaller fruit tortrix, codling moths, scab	all fruit pests, scab

- **Coloured traps:**
  These sticky traps can be used for the pests in Table 5.4.

Table 5.4. Types of traps.

Type of trap	Pests	Use
white trap yellow trap red trap	apple sawfly, plum sawfly cherry fruitfly	monitoring flight for monitoring in big orchards trapping and monitoring flight

- **Pheromone traps:**
  In these, the female's sex pheromone is emitted from a scent capsule and lures the male lepidoptera into an adhesive-coated wrap-around carton.

  Pheromone or sex traps are used for monitoring flight of the following pests: codling moth, smaller fruit tortrix, fruitlet mining tortrix, the tortricid moth *Capua reticulana*, *Archips podana*, *Pandemis heperana*, noctuid moths, winter moth, leopard moth, goat moth, leaf miners, apple tree clearwing moths, etc.

  The sex pheromones of the most important butterflies that cause damage to commercially grown fruit have been analysed and synthesized and are available from pesticide dealers.

Beneficials, or taking advantage of natural regulation

Organic fruit production relies first of all on the natural antagonistic relationships (relationship between the pest and beneficial), and pesticides etc. are used only if there is no longer any possibility of natural regulation. Some beneficials and their effects will therefore be briefly described.

Parasitoids (predatory parasites)

Hymenoptera

Hymenoptera are small wasps with a wasp waist. The larvae are like maggots and are either endoparasitic (living in the body of the host) or ectoparasitic (attached to the outside of the host animal).

The eggs are laid into or on to the host animal by means of an ovipositor. The larvae feed first on the less important organs of the host animal, so that the latter can remain alive until the parasitic larva pupates. The success of a parasitization is not established until the next generation.

The Hymenoptera are subdivided into three groups:

- ichneumon flies
- braconid wasps
- chalcidid wasps.

Fig. 5.2. Adult ichneumon fly, the female has a long ovipositor.

Fig. 5.3. Adult braconid wasp.

Fig. 5.4. *Prospaltella* leaving the shell of a San José scale.

Fig. 5.5. Hymenoptera larvae leaving a parasitized caterpillar pest.

ICHNEUMON FLIES (ICHNEUMONIDAE). The distinguishing feature of the ich-neumon flies is a mark on the wing. They are not very often found in orchards.

Ichneumon flies parasitize: leafroller caterpillars.

BRACONID WASPS (BRACCONIDAE). These are smaller than ichneumon flies (1–10 mm in size) and not so strikingly coloured.

Braconid wasps parasitize: leafroller caterpillars, codling moth caterpil-lars, leafminer caterpillars, mealy apple aphid, green apple aphid.

CHALCIDID WASPS (CHALCIDIDAE). These can be identified by their metallic lustre. Their wings have virtually no veining.

Chalcidid wasps parasitize: leafroller eggs and caterpillars, codling moth caterpillars, plum leafroller caterpillars, fruit tortricid caterpillars, San José scale, woolly aphids and small ermine moths.

Tachinids (Tachinidae)

Tachinids are similar to house flies, but they are highly adapted insect parasites. The adult flies are pollinators. The eggs are laid on the body of the pest. The caterpillars quickly bore into the body of the host animal. Some of the eggs are ingested in feeding.

Tachinids parasitize: winter moth caterpillars, codling moth caterpil-lars, noctuid moths, small ermine moths, beetle larvae.

The most important predators

Hoverflies (Syrphidae)

Adult hoverflies are pure pollinators and are distinguished by their typi-cal black-and-yellow colouring and characteristic flight behaviour. They can hover 'motionless' in the air and then shoot off like a rocket if alarmed.

The larvae lay their eggs individually in aphid and leafsucker colonies. The larvae are headless and are greenish, reddish and brown-ish in colour. They are voracious and are one of the most important aphid predators.

Hoverfly larvae feed on aphids and leafsucker larvae.

Gall midges (Cecidomyiidae)

Apart from the species which cause damage to plants, there are also a few useful species whose larvae feed on aphids.

Fig. 5.6. Adult hoverfly: this insect looks like a wasp and is an active pollinator.

Fig. 5.7. Hoverfly larva: transparent and pointed at the front, with no feet.

Beetles (Coleoptera)

There are over 5000 species of beetle in central Europe, some of which are useful and others harmful.

The most important useful species is the ladybird, but ground beetles, soft beetles, predatory beetles and rove beetles are also known enemies of pests.

Ladybirds (Coccinellidae)

These feed mainly on aphids, woolly aphids and scale insects, and also on some species of spider mites and mildew fungi.

Ladybirds have a body which is highly convex on top but flat underneath. They have a variegated colouring with red or yellow wing covers and dark spots, or dark wing covers with light-coloured patches or spots.

Ladybirds start to feed on aphids as soon as they appear in the spring. The rather long yellow eggs are laid in aphid colonies. The larvae have small nodules on top and stronger legs but weaker mouth parts than green lacewing larvae.

The most important ladybirds are:

- **seven-spot ladybird** (aphid predator)
- **two-spot ladybird** (aphid predator)
- **ten-spot ladybird** (aphid predator)
- **fourteen-patch ladybird** (aphid predator)
- **four-patch ladybird** (feeds on aphids, scale insects and woolly aphids)
- **fourteen-spot ladybird** (aphid predator)
- **twenty-two-spot ladybird** (feeds on mildew and sooty mould fungi)
- **dwarf ladybird** (feeds mainly on spider mites).

Fig. 5.8. Gall midge larvae (without feet, orange) feed on aphids.

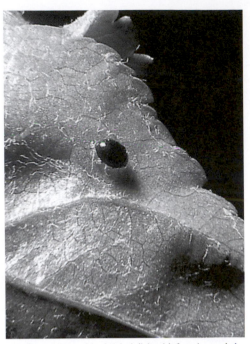

Fig. 5.9. Dwarf ladybird (black) feeds mainly on spider mites.

Fig. 5.10. Ladybird eggs, yellow in colour, are laid near aphid colonies.

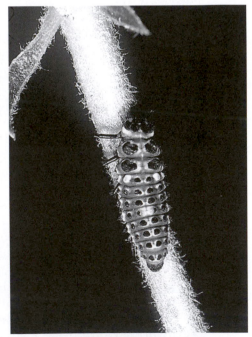

Fig. 5.11. Ladybird larva: has strong thoracic legs and moves very fast.

Green lacewings (Chrysopidae)

These come into contact with man very frequently, as they like to spend the winter in attics and living rooms.

The adult green lacewings have interleaved wings of a green to reddish brown colour.

The larvae have stronger mouth parts and weaker legs than ladybird larvae and feed on aphids, spider mites, gall midges, thrips and scale insects.

Bugs (Heteroptera)

Bugs have a flat body and a stiletto-like rostrum. There is a small triangular shield behind the conspicuously large neck shield.

Bugs can be both useful and harmful (causing deformities in fruit), and the dividing line is not always clear.

The three most important families are:

- anthocorid bugs
- mirid bugs
- nabid bugs.

ANTHOCORID BUGS (ANTHOCORIDAE).   These include the genera *Anthocoris* and *Orius*. Whereas bugs of the *Anthocoris* genus, which are larger in size, are the most important enemies of the pear leafsucker, the smaller *Orius* species feed on spider mites. The *Anthocoris* bugs also feed on aphids.

MIRID BUGS (MIRIDAE).   These are delicate, soft and usually brightly coloured. Whereas some species specialize in aphids, others feed on spider mites. These bugs can also cause damage, however, in the form of stunted fruit.

NABID BUGS (NABIDAE).   These are very big and can even overcome leafroller caterpillars. A characteristic feature of this type of bug is the curved rostrum, which is kept close to the body when not being used. Their preferred food is aphids and other small insects.

Predatory mites (Phytoseiidae)

Predatory mites are the most important beneficials in apple orchards. They are about 0.4 mm in size, long, drop-shaped and milky white to orange-red in colour. They tend to congregate around the central rib on the underside of the leaf.

Predatory mites can also be vegetarian, feeding on fungal spores, hyphae or pollen.

The hibernating females become active immediately after budburst, and can initially be found around the calyx node at flowering. After flowering they migrate to the underside of the leaf.

Predatory mites are highly voracious and one predatory mite per leaf can control 8–10 spider mites per leaf. Apart from the red spider mite, they also eat rust mites.

Predatory mites are highly sensitive to pesticides.

Predatory mites can be introduced into the orchard either via felt strips or via shoots.

Nematodes

Nematodes are tiny creatures which occur mainly as pests. Some of the species living in the soil, however, are also beneficial and parasitize leafroller and beetle larvae.

They penetrate via body orifices into the pests and secrete bacteria which lead to the death of the host.

Bacteria

The most important bacterium is *Bacillus thuringiensis*. This is ingested by caterpillars with their food, and an endotoxin that is active against insects is produced in the course of further development.

Caterpillars which have ingested the bacterium stop feeding 1 day later but do not die until after a few days.

There are three different pathotypes (strains), but only strain A is significant as a means of controlling caterpillars in orchards.

Bacillus thuringiensis is especially effective against winter moth caterpillars.

Viruses

Granulosis viruses are known to have a beneficial effect in fruit growing. They are ingested by caterpillars (codling moths and leafrollers) with their food and lead to the death of the pests.

Fungi

Fungi can attack and kill virtually all plants and animals.

The most important fungi that are useful in fruit growing are:

- *Beauveria* species (parasitize cockchafer larvae)
- *Coniothyrium* species (against San José scale)
- *Beauveria* species (against codling moths)
- *Athelia bombacina* (against apple scab).

Mammals

The most important mammal is the weasel. It is a highly effective predator, helping to control voles, but it is only found in remote, quiet locations with water nearby.

Fig. 5.12. Adult green lacewing: a beautiful insect with lacy wings and compound eyes.

Fig. 5.13. Eggs of the green lacewing on the characteristic stalk.

Fig. 5.14. Adult *Anthocoris nemoralis*.

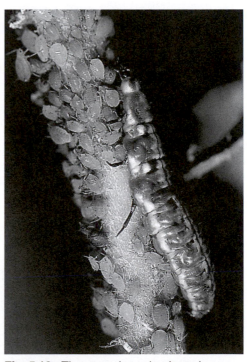

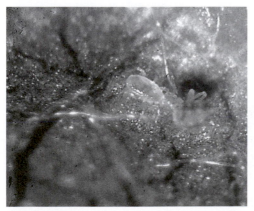

Fig. 5.15. Predatory mite feeding on a red spider mite.

Fig. 5.16. The green lacewing larva is equipped with two suckers and has 'weaker' legs than the ladybird larva.

Other mammals, such as hedgehogs, shrews, bats, cats and dogs, also contribute to the ecosystem, however.

Birds

Songbirds (coal-tit and blue-tit) make a valuable contribution to the biological control of leafroller caterpillars by feeding on these pests.
Birds of prey help to control mice.

Essential characteristics in beneficials

To be as effective as possible, beneficials should have the following characteristics:

- same life span as the pest
- high reproductive potential
- the ability to attack more than one species and eat large numbers of individuals
- the ability to find their prey, and thus survive, even when prey density is low
- they should be as resistant as possible to unfavourable climatic conditions.

Table 5.5. How many pests can a beneficial eliminate?

Beneficials	Pests eliminated per day	Total number of pests eliminated
Predatory mites	5 spider mites	30–50 spider mites
Predatory bugs (*Orius* species)	30 spider mites	200 spider mites
Globe beetles	30 spider mites	250 spider mites
Ladybirds	10–50 aphids	400 aphids
Green lacewing	30–50 spider mites	200–500 aphids
Hoverfly	10–40 aphids	150–160 aphids
Woolly aphid predator wasp		up to 90% parasitization in autumn
San José scale predator wasp		70–90% parasitization in autumn
Aphid predator wasp		200–1000 aphids

Major diseases and pests of pome fruit

Diseases

Apple scab (Venturia inaequalis)

NATURE OF DAMAGE. First of all, small oily blotches can be observed on the young leaves. The typical ramifications can be seen in these blotches with a magnifying glass. The blotches subsequently turn olive-green to black in colour. In summer the centre of the blotch can die off and fall out, giving the leaf the appearance of being riddled by small shot. Severely affected leaves drop off.

Brownish black patches are found on the fruit. If the infection is only slight, these patches become corky, but if the infection is severe, cracks appear on the fruit, which starts to go rotten. If the fruit is attacked at a very early stage it may develop deformities. The later the infection occurs, the fewer are the cracks and the smaller the patches formed.

If infection occurs just before the harvest, the symptoms do not develop until the apples are in storage (storage scab), in the form of small black spots.

If vigorously growing trees are severely attacked by scab in the autumn, so-called shoot base scab may occur in the following spring. In this condition the scab mycelium grows out with the new shoot and attacks the shoot at its base (particularly common in Golden Delicious). The mycelium grows to a length of about 3–10 cm.

Fig. 5.17. Leaf scab: dark spots which go on to produce spores.

Fig. 5.18. Very severe fruit scab.

VARIETAL SUSCEPTIBILITY. The following varieties are highly susceptible to scab: Arlet, Gala, Golden Delicious, McIntosh, Gloster, Rubinette, Summerred, Jonagold and Braeburn.

Under organic conditions it is preferable to grow varieties that are robust or resistant to scab (see section in Chapter 3 on *Apple cultivars*).

As one form of resistance has already been broken, however, a minimal plant protection programme against scab should be implemented as a preventive measure even with these resistant varieties.

COURSE OF THE DISEASE.   The scab fungus has two possibilities for overwintering. Usually it overwinters in fallen leaves, but it may also overwinter in buds and on shoots if the infection is severe, tree growth is
vigorous and ideal infection conditions are present in late summer and
autumn.

PRIMARY INFECTIONS.   In the saprophytic phase (sexual reproduction) the
development of the fungus starts with the fall of the leaves in autumn.
The hyphae start to proliferate through the entire leaf and grow towards
each other. Sex organs are developed and rudimentary perithecia are
formed. After a dormant phase in the winter, the perithecia and the
ascospores which they contain start to mature in the spring. Maturation
is accelerated by warm wet weather, and slowed down by dry or cold
weather.
   Some ascospores are already mature at budburst, and within the next
6–10 weeks all of them become mature and are ejected into the air. The
maximum number of mature spores is usually attained around flowering
time.
   For the ascospores to be ejected, the following conditions must be
present:

- The ascospores are only ejected when it is raining (a mere 0.2 mm of
  rainfall is sufficient, however).
- The ascospores need light in order to be ejected. Not more than 5%
  of the mature spores are ejected at night.
- Temperatures below 4°C can greatly suppress the scattering of the
  ascospores.
- Brief rainfall is not sufficient to eject the spores after long dry periods.
- The number of spores ejected depends on the specific conditions in
  the orchard and the proportion of mature and ejectable spores at the
  time.

   The ascospores ejected are carried to the leaves by the wind. When
the leaf is wet the spores start to germinate, and they can penetrate into
the leaf if it stays wet for long enough. Penetration into the leaf takes
place very easily if the leaf is young, but older leaves become resistant to
the fungal spores (age resistance).
   As in the maturation of the spores, not all the spores are at the same
stage of development when they germinate and penetrate into the leaves;
some individuals are faster and others slower.
   Germination of the spores stops if the leaf becomes dry for a time,
and if the dry phase is prolonged the first spores start to die off. If the
leaf gets wet again before all the spores have died off, the spores can continue to germinate and cause infection.
   As soon as the spores have penetrated into the leaf they are no longer
dependent on leaf wetness. The mycelium now spreads under the leaf
surface without penetrating deeper into the leaf.

DEGREE OF PRIMARY INFECTION.   The degree of scab infection depends on the following factors:

- initial level of infection (infection the previous autumn)
- putrefaction of fallen leaves
- proportion of mature, ejectable spores
- proportion of young, susceptible leaves
- susceptibility of the cultivar
- weather conditions.

MILLS OR RIMPRO?   For decades the Mills scab table was considered to be the most important tool for calculation of infections. Large numbers of unsuccessful attempts at scab control, however, have cast doubt on the accuracy of this table, which only takes into account the physical factors relevant for infection. It ignores the biological factors (e.g. spore potential, spore maturity, leaf growth), and it is questionable whether the calculations are correct.

Marc Trapman (NL) has developed a scab simulation program called 'RIMpro', which takes into account the biological factors as well as the physical ones.

On the basis of meteorological data for any given point in time, RIMpro calculates the percentages of spores which are immature, mature, ejectable, ejected or germinating, or have penetrated into the leaves, and can thus provide an up-to-date picture of the progress of primary infections.

This program is of particular importance for organic apple orchards, as it is the only program that can also be used for the preventive application of fungicides.

Table 5.6. Factors affecting scab infection.

The severity of scab infection is increased by	The severity of scab infection is reduced by
high initial infection in the autumn	low initial infection in autumn
slight or slow putrefaction of fallen leaves in the spring	rapid or almost complete putrefaction of leaves in the spring
large numbers of mature and ejectable spores	few ejectable spores
periods of warm wet weather (temp. 10–15°C)	cold or dry weather
vigorous leaf growth as a result of high-temperature-susceptible cultivars	little leaf growth
	resistant cultivars

SECONDARY INFECTIONS.   Soon after the appearance of the first symptoms (about 2–3 weeks after infection) the conidiophores break through the cuticles and form the pear-shaped summer spores (conidia). These are produced in large numbers and can lead to an explosion of scab infection if weather conditions are optimal. Conidia are released and spread by raindrops and also by wind. This means that the infection progresses with the movement of the raindrops (i.e. from tree to tree). Unlike the ascospores, therefore, secondary infections do not need rain or light to spread, and so even long periods of dew in the night can lead to secondary infections by conidia.

In scab-susceptible cultivars the development of the secondary infection is much faster and more explosive than in robust cultivars.

When primary infections have developed, secondary infections can occur until shoot production ends. The fungus overwinters in the fallen leaves.

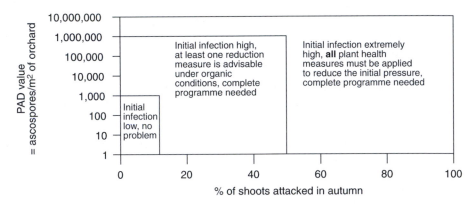

Fig. 5.19. Relationship between autumn infection and control programme.

OVERWINTERING ON THE TREE. If there is severe secondary infection and trees are producing shoots, infections may develop in buds and shoots. These overwintering forms on the tree have already produced large numbers of conidia by budburst and can thus lead to early infections of the

Table 5.7. Differences between primary infections and secondary infections.

Type of infection	Primary infections	Secondary infections
caused by	ascospores	conidia
how spores are spread	ejection of spores	release of spores
conditions required for ejection or release of spores	rain, light (not more than 5% of spores ejected at night), high temperature at the time of ejection	wind or rain (but not necessary)
number of patches/leaf if infection occurs	usually only 1–2 patches	many patches
effect of total hours of leaf wetness on the infection	moderate	very great

sepals. These infections can later be seen on the fruit in the form of scab patches, radiating from the calyx.

REDUCTION OF THE BASELINE POTENTIAL. The baseline potential (the level of infection in the autumn) is the critical factor which determines the course of the disease in the next spring. The number of spores in the coming spring can very easily be inferred from the percentage of infected shoots in the autumn.

If less than 10% of the shoots are infected in the autumn, the baseline potential is very low and it is possible to achieve effective control of scab next spring using the fungicides available.

If up to 50% of the shoots are infected, the baseline potential will be very high, and in an organic orchard it is then advisable to reduce the level of spores for the next year. Even individual measures have a positive effect here.

If more than 50% of the shoots are infected in the autumn, all possible measures to reduce the number of spores must be applied in order to bring the scab under control again. Individual measures to reduce the spore potential are no longer sufficient.

The following measures are available in organic orchards for reducing the baseline potential:

SHREDDING THE FALLEN LEAVES. The fallen leaves can be shredded in the autumn or in the spring. Microbial degradation is accelerated by tearing up the leaves. In extreme situations shredding should be carried out both in autumn and in spring.

Effectiveness: 90% reduction in the spore count, 50% reduction in infection.

ROTAVATING THE FALLEN LEAVES INTO THE SOIL. The line of trees is often rotavated in autumn in order to reduce the problem of voles. The leaves lying between the trees are also rotavated into the soil in this operation. The leaves in the tramline, however, are ignored.

Effectiveness: 50% reduction in the spore count.

CONSERVATION OF EARTHWORMS IN ORDER TO ACHIEVE FASTER BREAKDOWN OF LEAVES. The earthworm is the most important scavenger of fallen leaves and thus the most important ally in the fight against the scab fungus. Earthworm density is normally very high in organic orchards. Copper sprays and mechanical soil tillage can have a negative effect on earthworm populations, however. Care should therefore be taken to ensure that the amount of copper spray used is as small as possible.

Effectiveness: 90% reduction in spore production.

USE OF ANTAGONISTIC FUNGI. Some fungi, such as *Athelia bombacina*, are hyperparasites of the scab fungus, i.e. they kill the scab fungus in the fallen leaves. These fungi are very easy to grow on wheat bran and can be sprayed on to the leaves before they fall in the autumn. This method is not yet sufficiently developed for commercial application, however.

All these plant health measures are of no real use, however, unless the shoots of the trees stop growing early in the autumn and the over-wintering of the scab fungus on the tree and the production of conidia can be prevented.

CONTROL

PREVENTIVE

- Do not plant cultivars that are susceptible to scab.
- Choose suitable sites. In the summer, secondary infection develops much faster in sites where leaves stay wet for a long time (because of dew).
- Apply suitable cultural measures to make sure that the shoots stop growing early.
- Mixed plantings can significantly slow down the development of infection in the secondary season (see section in Chapter 3 on *Choice of cultivars*).
- Carry out plant health measures.

DIRECT.   Scab is the most important disease in fruit production, and more than half of the measures required are for the purpose of scab control. Unlike animal pests, scab is an 'invisible enemy', which does not become visible until the fruit tree is already infected. For this reason all treatments must be performed before an infection.

So as to be able to implement control that does not harm the environment and is specific to the orchard (and to the infection), the baseline potential must first be determined in the autumn. A strategy can then be chosen on the basis of the infection pressure that is specific to the orchard.

Successful scab control cannot be achieved unless all the people involved (growers, extension officers, etc.) act with the utmost flexibility. For the grower this means spraying more frequently (twice a week if necessary) at times when there is a risk of scab, and reducing spraying frequency at times when there is no scab risk.

The job of the extension service is to provide the grower with information about spore maturity, spore ejection and spore penetration as fast as possible, so that the grower can respond immediately.

Apple powdery mildew *(Podosphaera leucotricha)*

NATURE OF DAMAGE.   **In winter:** poor closure of the bud scales, the buds are often wrinkled and die off; infected shoots are whitish (powdery coating).

Table 5.8. Scab strategies.

Measure/time	Orchards with low baseline potential (<10% infected shoots) or resistant cultivars	Orchards with a high baseline potential (10–15% infected shoots)	Orchards with an extremely high baseline potential (>50% infected shoots)
shredding fallen leaves in autumn	not necessary	necessary	necessary
shredding fallen leaves in spring	not necessary	not necessary if leaves were shredded in autumn	necessary
in the greentip stage, if a period of bad weather is forecast		preventive spraying with copper	preventive spraying with copper
mouse ear stage – end of flowering: if a period of bad weather is forecast	preventive spraying with copper or with sulphur or calcium sulphide	preventive spraying with copper or with sulphur or calcium sulphide	preventive spraying with copper or with sulphur or calcium sulphide
if many spores have been ejected	spraying on wet leaf only if there has been no preventive spraying	spraying on wet leaf on the germinating spores	preventive spraying with copper or with sulphur or calcium sulphide
and if the preventive spraying was done 3–4 days earlier and a new leaf has already grown		spraying on the germinating spores	spraying on the germinating spores
secondary season: if scab infection is found in the orchard	preventive spraying: frequency depends on leaf growth and how long the leaves are wet	preventive spraying: frequency depends on leaf growth and how long the leaves are wet	preventive spraying: frequency depends on leaf growth and how long the leaves are wet

In spring: infected buds open late, the new leaves are covered with a white powdery deposit. They grow upwards at a steep angle and curl up from the edge on the top of the leaf, staying too narrow and becoming hard and brittle.

The underside of the leaf shows an initially reddish tinge, which later changes to a brown colour.

During the growing season: if the infection progresses, leaves are continuously shed, often leaving only a few dried leaves at the tip of the bare and stunted shoots. Infected shoots are shiny white.

Blossoms infected by mildew wither and are also covered with a powdery mycelium.

Fig. 5.20. Mildewed shoots in winter: typically there is a powdery layer with unenclosed apical buds.

Infected fruits remain too small in size and show a typical criss-cross russeting.

SUSCEPTIBILITY. Jonagold, Idared and McIntosh are highly susceptible varieties.

COURSE OF THE DISEASE

PRIMARY INFECTIONS. The mildew fungus overwinters as mycelium in the buds. If there is very severe frost in winter, some of the infected buds may die off, as the bud scales are not completely closed. In the closed buds the mycelium produces conidia, which infect the young leaves, blossoms and fruit when the buds open. Infection may occur at the greenbud or redbud stage if weather conditions are favourable. The first symptoms of primary infections can then be seen on the leaves and petals at flowering. Apart from the climatic conditions, for mildew infection to develop the tree must also have susceptible tissue (very young leaves) available. The leaves acquire age resistance to mildew at an even earlier stage than they do to scab.

Mildew infections peak during the period of greatest shoot growth (around flowering). In this period there is also often infection (and consequently also russeting) of the fruit.

SECONDARY INFECTIONS. Infected leaves produce conidia which can in turn infect young, susceptible leaves. The fungus develops first on the underside of the leaf. In autumn the fungus grows down the stem and penetrates into the buds which are forming, before the bud scales become corky. This is also a particularly critical point in mildew infections, as both lateral and terminal buds can be infected.

WEATHER CONDITIONS FOR A MILDEW INFECTION. Optimum weather conditions are an air temperature of 15–25°C and over 70% relative humidity.

At temperatures of 4–10°C the germination of the mildew spores is very slow.

Sultry, warm and wet periods, often after rain, are optimum conditions for infection.

Mildew infections cannot occur when the leaf is wet.

CONTROL

PREVENTIVE

- Remove mildew-infected shoots during winter pruning and late summer pruning. This greatly reduces the spore potential. If there is severe infection, however, pruning should not be done at a time when there is vigorous shoot growth, since the shoots then start to sprout again and new, young and thus susceptible leaves are formed. Anti-mildew pruning at the wrong time can increase the infection.
- The period of infection is reduced if the trees finish shoot production as early as possible.
- Plant varieties that are resistant or have low susceptibility to mildew.

DIRECT. Anti-mildew spraying is only necessary when varieties are susceptible and there is moderate infection pressure. In other cases the preventive measures are sufficient.

Sulphur chemicals (3 kg/ha) are used for control of mildew. Control must start at the green–redbud stage and continue until the end of shoot production.

Sulphur has a preventive effect on spore germination.

Spraying frequency depends on leaf growth. In the periods of most vigorous shoot growth, when 2–3 new leaves are produced every week, the spraying frequency needs to be increased to once every 4–7 days. When there is little growth the frequency can be reduced to once every 10–14 days.

Sulphur should not be sprayed at temperatures over 30°C, as it may cause scorching.

Apple canker (Nectria galligena)

NATURE OF DAMAGE.   Small, reddish brown depressions are found on the shoots, usually around an eye. When the site of the infection gets bigger, the bark peels away and cankerous lesions develop (open canker). The shoots above the site of the infection gradually die. Larger canker sites, which the tree tries to close with occlusion burr knots, develop on older branches and on the trunk (closed canker).

Both whitish conidial deposits and reddish perithecia can be found on infected shoots.

SUSCEPTIBILITY.   The varieties Gala, Cox's Orange Pippin, Gloster, Idared, Summerred and McIntosh are highly susceptible to apple canker.

Fig. 5.21. Canker growth on the stem.

Fig. 5.22. Mildewed shoot in summer: leaves with a white powdery covering layer.

COURSE OF THE DISEASE. Dispersion of the ascospores of the fungus starts before autumn. The spores are spread by rain and can penetrate into the tree via wounds. The main entry points in autumn are the fruit and leaf scars. Dispersion of the ascospores continues in the spring, and there is also increased production of conidia on infected shoots in the second half of the year. Production of conidia continues until late autumn. The best conditions for infection are thus in autumn, when there are a lot of wounds and spore production peaks.

Infections can also occur in spring and summer via wounds caused by pruning, hail, frost, etc., but cannot occur unless there is some form of wound.

CONTROL

PREVENTIVE

- Do not plant susceptible cultivars.
- The trees should finish shoot production as early as possible, to ensure that the leaf-drop period is uniform and short. The longer this period lasts, the greater is the possibility of infection.
- Inspect young trees carefully when they are planted, to make sure that no canker-infected trees are planted.
- Make sure that trees have good vitality. Healthy trees are less likely to become infected.
- Make sure the soil is not waterlogged. Trees on wet soils are more prone to developing canker.
- Make frequent checks for canker infection, especially in young orchards.
- Diseased trees should be removed immediately from the orchard in the first 2 years after planting.

- Infected prunings should be removed from the orchard and burnt.

DIRECT

- Cankered areas should be cut out during the winter and covered with a wound sealer.
- If there is infection, the tree should be treated with copper after the harvest, at the start of leaf drop, during leaf drop and after pruning, when a period of rain is forecast.

Flyspeck and sooty blotch

Table 5.9. Causes and damage.

	Flyspeck	Sooty blotch
Causative organism	*Schizothyrium pomi*	*Gloeodes pomigena*
Damage	dark spots on the skin of the fruit, often in groups	blotches on the skin of the fruit, which can be washed off

COURSE OF THE DISEASE. The two diseases often occur together. The causative organisms overwinter on apple trees and other broad-leaved trees. Infections occur from the end of May onwards, in cool, wet weather. There is a risk of infection until September. Severely affected apples start to shrivel in storage.

CONTROL. In old orchards these two fungal diseases are the second most important fungal disease after scab. In some years the infection can affect the entire crop in densely planted orchards.

PREVENTIVE. Make sure the crown is kept open. Pruning and fertilizer application should be optimized. Infection is much worse in old trees and in trees with dense crowns than in young trees.

Remove host plants (ash, sycamore, lime and willow) at the edge of the orchard.

DIRECT. Repeated spraying with coconut soap from June onwards can significantly reduce infection.

Calcium sulphide is also reported to have a slight effect on both these diseases.

Fruit rots and storage rots (Monilia, Gloeosporium rot, etc.)

The organisms that cause these rots penetrate into the fruit either via wounds (*Monilia*, etc.) or via the lenticels (*Gloeosporium* rot, *Phoma*, etc.). They overwinter in mummified fruit, prunings, fallen fruit, fallen leaves and other woody plants.

In fungal diseases the entire fruit rots if the causative organism enters via wounds. If it enters via lenticels, it produces patches 2–5 cm in size.

CONTROL OF FRUIT ROTS AND STORAGE ROTS

PREVENTIVE

- Remove mummified fruit and other sources of spores from the tree.
- Make sure there is good soil life, so that fallen leaves, prunings and fruit lying on the ground are broken down (in particular take care not to harm earthworms).
- Remove host plants at the edge of the orchard (especially ash and other broad-leaved trees).
- Do not harvest wet fruit for storage.
- Optimize the timing of the harvest.
- Place the harvested apples in storage immediately and establish the right air composition and storage temperature.

DIRECT. There are no direct measures.

Pear rust (Gymnosporangium fuscum)

NATURE OF DAMAGE. Orange-red patches, up to 1 cm in size, develop on the top of the leaves in spring (June). Red, gristly pimples form on the underside of the leaf in the course of the summer.

COURSE OF THE DISEASE. The causative organism of pear rust is a fungus which changes host, i.e. it undergoes part of its development on other plants (juniper).

In spring the fungus first attacks the juniper bush. In wet weather the spore deposits on the juniper trees can swell up and get on to the pear tree leaves, where they penetrate and produce orange-red patches on the top after 3–4 weeks. The spore deposits are formed on the underside of the leaf. These spores return to the juniper trees, where the pathogen overwinters.

The greatest risk of infection is within a radius of 30–50 metres from juniper bushes.

Severe infection every year reduces assimilative performance and leads to a decline in productivity and eventually to the death of the tree.

CONTROL

PREVENTIVE

- The only control measure which has been found effective is to remove all juniper bushes within a radius of 30–50 metres from the pear trees.
- New orchards should not be planted near existing juniper bushes.

DIRECT. There are no direct control measures.

Pests

Red spider mite (**Panonychus ulmi***)*

APPEARANCE. **Adult**: Females about 0.4 mm long, body oval, highly convex and dark red with long bristles; short legs, light in colour. The yellowish green to bright-red males are somewhat smaller and more or less pear-shaped.

Egg: 0.17 mm diameter, dark red, onion-shaped.

Evolutionary stages: the larvae have six legs and the nymphs eight legs, yellowish green to bright-red. The red spider mite overwinters as a red winter egg mainly on the branch forks of 2- and 3-year-old wood. If the infestation with red spider mite is very severe, the winter eggs can also be found on 1-year-old wood. The eggs begin to hatch at about the redbud stage. The young larvae migrate immediately to the rosette leaves and start to suck them. Hatching continues in May, when the larvae which hatched first have already completed their development. As the temperature rises, development becomes very rapid, with 5–7 overlapping generations. The somewhat paler summer eggs are laid on the underside of the leaf. The females start to lay winter eggs and stop reproduction when night-time temperatures drop and the days get shorter.

Massive increases in red spider mite occur either in hot dry summers (when conditions for reproduction of red spider mite are ideal) or sometimes after wet periods (when the predatory mites are harmed by the frequent spraying of sulphur-based fungicides to control scab).

LIFE CYCLE

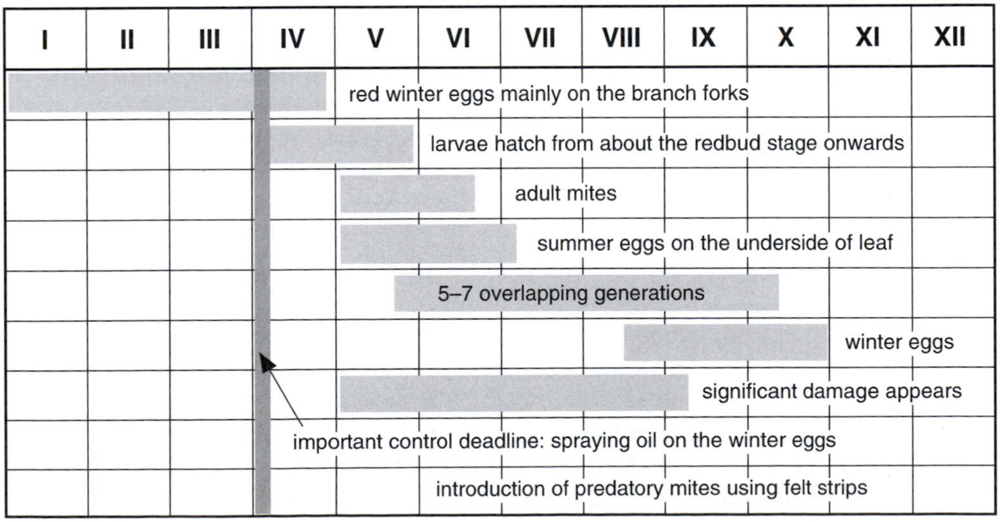

I	II	III	IV	V	VI	VII	VIII	IX	X	XI	XII
				red winter eggs mainly on the branch forks							
				larvae hatch from about the redbud stage onwards							
				adult mites							
				summer eggs on the underside of leaf							
				5–7 overlapping generations							
				winter eggs							
				significant damage appears							
			important control deadline: spraying oil on the winter eggs								
			introduction of predatory mites using felt strips								

Fig. 5.23. Life cycle of red spider mite (*Panonychus ulmi*).

NATURE OF DAMAGE.   The first sign is slight speckling of the leaves due to sucking by the mites. If infestation is severe the leaves become dull green, brownish and eventually bronze-coloured and brittle. Some leaves drop off. This has a negative effect on fruit quality and flowerbud production. The fruit fail to develop sufficient colour (Golden Delicious, for example, stays green) and flowerbud production is inhibited. Cropping capacity for the next year is thus lower.

ENEMIES.   Under natural conditions the red spider mite has very many enemies and should therefore not be a major problem in organic production. By far the most important enemy is the predatory mite, but globe beetles, green lacewings and the anthocorid bug *Orius minutus* can also provide valuable assistance. Depending on weather conditions, in some years these predators can achieve control of spider mites even without the aid of predatory mites.

As beneficials, predatory mites have the big advantage that they stay on the tree the whole year round and are thus always there when spider mites are present (protective predators).

These beneficials have a drop-shaped body and are milky white to orange-red in colour (if they have just eaten a red spider mite). They overwinter as adult females in cracks on the trunk and emerge in the spring before flowering. They can be found initially in the down of the sepals and in the calyx node, but later they congregate on the underside of the leaves, where they prefer the area around the central rib. Predatory mites like cultivars with highly pilose leaves (e.g. Jonagold and Idared) best of all. On less pilose leaves, as in Golden Delicious, it is much more difficult to use predatory mites effectively.

It is not always possible to achieve a stable balance between red spider mites and predatory mites immediately.

One predatory mite per leaf can control 8–10 red spider mites/leaf.

Predatory mites are highly sensitive, however, and are inhibited or killed by various pesticides. This should not really be a problem under organic conditions, but sulphur is often sprayed in organic orchards, where it is one of the most important pest control agents. Frequent spraying of high doses of sulphur, however, can cause severe damage to predatory mites.

Only with predatory mites is it possible to achieve lasting and problem-free control of the red spider mite.

Apart from predatory mites, the globe beetle, anthocorid bug *Orius minutus*, green lacewing and ladybird larvae are also highly effective allies against the red spider mite. Their disadvantage is that they are scavengers and do not come into the orchard until it is infested with red spider mite. By that time, however, the population of spider mites has usually already risen to such an extent that these beneficials can no longer keep them under control.

CONTROL

PREVENTIVE

- The most important thing to do is to conserve beneficials, especially the predatory mite.
- In scab and mildew control, care must be taken to ensure that the number of times the trees are sprayed with sulphur-based fungicides and the average quantities per hectare are kept as low as possible.

DIRECT. Predatory mites should be introduced if none are present. This can be done in two ways:

- **Introduction of predatory mites by means of felt strips:**
  Felt strips or coir cords are tied around the trunks of the donor trees in late summer, to induce the migrating females to use them as their winter retreat. These strips or cords are removed from the donor trees and hung on the recipient trees in winter or early spring.
  The donor trees must of course have a suitably high population of predatory mites in the summer.
  This method is very simple and can be carried out at a time when there is little work to do.
  Moreover, with this method predatory mites are ready to be brought into the orchard at budburst.
- **Introduction of predatory mites by means of water shoots in summer:**
  Shoots occupied by predatory mites are cut from the donor trees and transported in a pallet box to the recipient orchard, where the shoots are placed on the recipient trees. On the next day the predatory mites will migrate to the recipient tree.
  Care should be taken to ensure that at least one shoot is allocated to every tree and that the shoots are introduced according to the wind direction.
  In this method, which should be used in summer, the predatory mites can still produce up to one generation by the autumn. This means that the initial population for the following spring is very low, and in the following year it is often a very long time before the predatory mites become visible.

Under organic conditions, chemical control of the red spider mite is only possible in relation to the overwintering eggs. Mineral oil and rapeseed oil products are permitted for this purpose. Mineral oil is much more effective than rapeseed oil, however.

The official damage threshold in winter is 2000 winter eggs/m^2 of fruiting wood. For the grower, however, it is virtually impossible to determine this threshold. A simple method is to run the finger along the branches where they fork. If the finger is stained red, this means that a large number of winter eggs have been laid and spraying with oil is necessary.

With regard to spraying to control red spider mite, the following points should be borne in mind:

- Spraying can be carried out until the redbud stage.
- Spraying is more effective the closer it is to the hatching of the larvae from the winter eggs.
- When spraying, care should be taken to ensure that the crotch angles are completely covered by the oil. For this reason the sprayer should be moved along each row of trees first in one direction and then in the other. Fine spraying is more effective than normal spraying for covering the crotch angles with oil.
- There should not be any frost in the 2–3 days after the tree is sprayed with oil – otherwise there may be adverse effects (ranging from scorching of leaves to shedding of blossoms).
- Susceptible cultivars, such as Gala, Cox's Orange Pippin, Boskoop, Braeburn and Rubinette, must be treated at the mouse-ear stage.
- The amount of oil used depends on the phenological development:
  - up to the mouse-ear stage: 30 l/ha
  - at the greenbud stage: 20 l/ha
  - at the redbud stage: 10 l/ha.

Apple rust mite *(Aculus schlechtendali)*

APPEARANCE. **Summer females** (protogynes): 0.16–0.18 mm in length, body yellowish brown with tiny nodules.
Winter females (deutogynes): similar, but with fewer nodules.

LIFE CYCLE

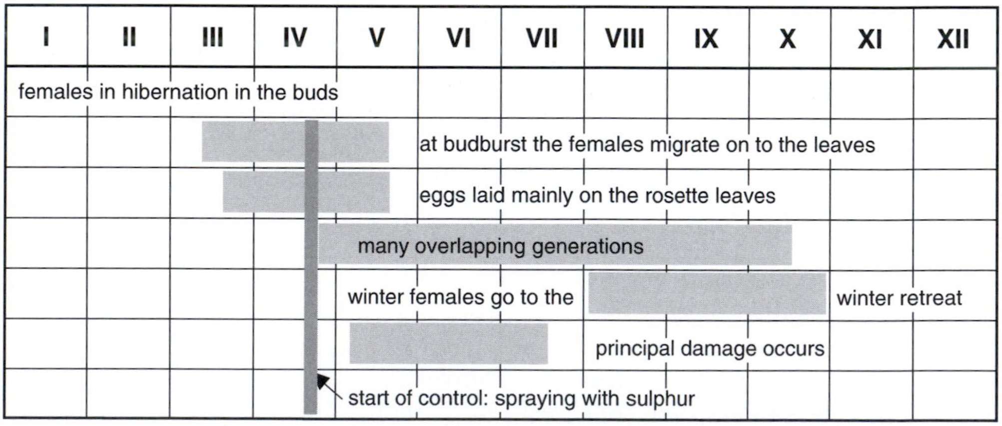

I	II	III	IV	V	VI	VII	VIII	IX	X	XI	XII
females in hibernation in the buds											
					at budburst the females migrate on to the leaves						
					eggs laid mainly on the rosette leaves						
				many overlapping generations							
				winter females go to the					winter retreat		
						principal damage occurs					
					start of control: spraying with sulphur						

Fig. 5.24. Life cycle of apple rust mite (*Aculus schlechtendali*).

The deutogynes (winter females) overwinter under bud scales, in cracks in the bark near bud scales, or under moss and algae on the bark. As soon as the buds open, the overwintering mites migrate to the rosette leaves and immediately begin to suck on them and on the blossoms. By the mouse-ear stage all the mites have migrated from the winter retreats.

In May the males and the protogynes (summer form) of the first generation hatch, followed by a number of overlapping generations. Formation of the winter forms, which move to their winter retreats from August onwards, starts at the end of June.

There is often a big increase in rust mite infestation from mid-May to mid-June, when the weather is very warm and dry, to such an extent that the proportion of leaves occupied by rust mites can increase from 10% to 100% within a week.

NATURE OF DAMAGE. The upper side of the leaf has a speckled, pale and dull appearance as a result of sucking by the mites. The underside of the leaf has a brownish discoloration. More severely affected leaves become silvery and can eventually turn brown. They then curl up from the edge inwards.

The damage to the leaves results in inadequate colour development in the fruit. Cultivars with a red top colour (e.g. Jonagold) are especially affected. This adverse effect is more common than the russeting caused by this mite.

There is a risk of russeting due to rust mites if there is severe rust mite infestation at or shortly after flowering.

ENEMIES. The most important enemy of the apple rust mite is the predatory mite. But whereas control of red spider mite is achieved very quickly with the predatory mite, in the case of the apple rust mite it often takes 1–2 years longer to reach a stable situation.

DAMAGE THRESHOLD. Damage thresholds usually relate to the number of rust mites per leaf. For the grower, however, this figure is impossible to determine. It is therefore simpler to use the infestation index method.

According to Marc Trapman (NL), there is a stable situation if not more than 20% of the leaves are infested with rust mites (6 out of 30 leaves). If rust mite infestation is greater than this, it is often no longer possible for predatory mites to restore the balance.

CONTROL

PREVENTIVE. Conservation of predatory mites.

DIRECT

- Introduction of predatory mites (see section on *Control of the red spider mite*).
- Chemical control:
  Spraying with sulphur from the mouse-ear/greenbud stage onwards has a good suppressive effect on the apple rust mite. Spraying with sulphur is less effective once infestation is very severe, however.

The amounts used should not be greater than 3 kg/ha, so as not to harm the predatory mites.

In orchards with rust mite problems, spraying with sulphur should be continued until the middle or end of July.

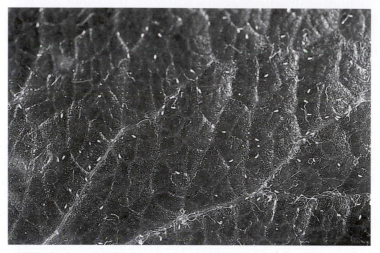

Fig. 5.25. Apple rust mite: small, comma-shaped, orange mites on the underside of the leaf.

Rosy apple aphid *(Dysaphis plantaginea)*

APPEARANCE. **Aphid**: initially dark green, then green with a rather powdery coating, then green with a little violet at the end of the body, then grey-violet and finally black with a powdery coating.
Egg: 0.6 mm, black, very difficult to distinguish from other aphid eggs.

LIFE CYCLE

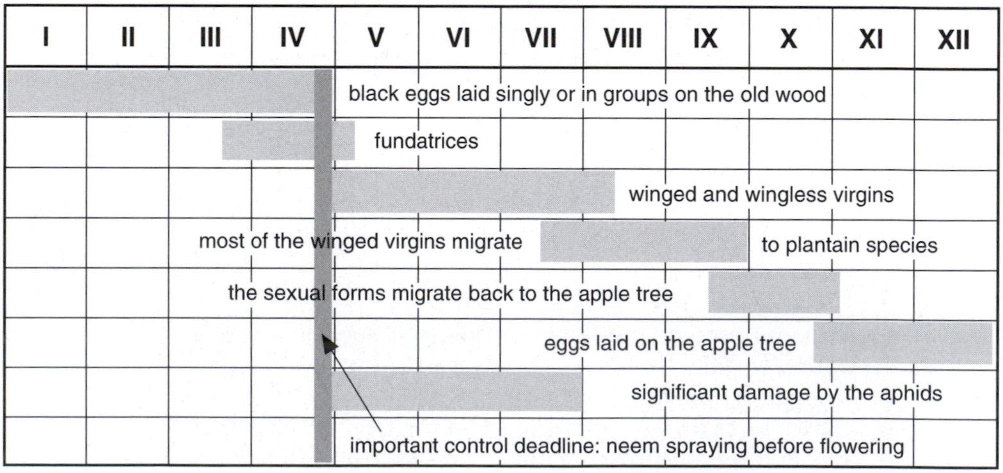

I	II	III	IV	V	VI	VII	VIII	IX	X	XI	XII
				black eggs laid singly or in groups on the old wood							
			fundatrices								
					winged and wingless virgins						
		most of the winged virgins migrate						to plantain species			
			the sexual forms migrate back to the apple tree								
					eggs laid on the apple tree						
					significant damage by the aphids						
			important control deadline: neem spraying before flowering								

Fig. 5.26. Life cycle of rosy apple aphid (*Dysaphis plantaginea*).

The rosy apple aphid overwinters as black winter eggs singly or in groups on the old wood. The first fundatrices hatch from their eggs at budburst and make a fundatrix gall, i.e. they turn the tip or edge of the leaf down and roll it up. These galls can be found only with great difficulty in the mouse-ear/greenbud stage. After about 2–3 weeks the fundatrices are adult and start to produce their parthenogenetic progeny. After flowering, the first-generation aphids themselves start to produce offspring. The infestation can increase very rapidly at this time.

The reproduction rate of the rosy apple aphid is enormously high, and so it can cause very great damage. Apart from the distinctive rolling of the leaves, the fruits are also damaged by the sucking of the aphids. The fruits stay small and deformed and do not drop.

This damage to the fruit can occur during or shortly after flowering. The shoots are rolled up (curled tail shoots). Shoots attacked by rosy apple aphids do not usually develop flower buds.

From the beginning of June onwards (when the embryo is about 15 mm in size) some of the aphids acquire wings and migrate to the plantain (the summer host). If the vegetation is very lush, however, some of the rosy apple aphids stay on the apple tree. The wingless aphids are spread by ants and the wind to other apple trees and produce new colonies there. The winged aphids migrate immediately to the plantain. New colonies can arise in this way until late summer, but they are not as dangerous as those produced in the spring.

In the autumn, winged sexual forms develop. These return to the apple tree if they have not remained there all the time. The fertilized females then lay their winter eggs.

Fig. 5.27. Rosy apple aphid: severely affected shoot.

ENEMIES.   Aphids have many enemies, but in the case of the rosy apple aphid they are not effective enough, especially in spring. **The most important enemies are**:

- green lacewings
- ladybirds
- ichneumon and chalcid wasps

- anthocorid bugs
- gall midges
- hoverflies.

These beneficials have a disadvantage relative to the rosy apple aphid: they can only reproduce very slowly in cold spring weather.

When the rosy apple aphid subsequently starts to produce nests, from June onwards, the conditions for development of the beneficials are better and they are able to keep the aphid populations in check. The beneficials thus play a significant role mainly in summer, but for the reasons just explained they are less important in the spring.

INSPECTION AND DAMAGE THRESHOLD.   Because of the rapid reproduction rate and the enormous damage that the rosy apple aphid can cause, the damage threshold is very low.

1. Visual inspection of the rosette leaves for fundatrices before flowering (green–redbud stage): damage threshold: 1% infestation.
2. Visual inspection after flowering (rosette leaves) and during the summer (long shoots): damage threshold: 1–3%.

CONTROL

PREVENTIVE

- Avoid rampant growth of trees
- conserve beneficials
- grease-banding to prevent ants getting on to the apple trees. The development of the aphids takes a different course when no ants are present.

DIRECT.   Control with natural enemies is almost impossible before and around flowering. Chemical treatment should therefore be carried out at this early stage.

Pre-flowering treatment with neem formulations (according to the latest EU directive, these can only be used in nurseries) – these sprays must be used before flowering, as neem acts very slowly.

Where neem is not permitted, additives can be used (wetting agents or soap formulations). These are not as effective, however. Pyrethrum–rotenone formulations are also ineffective.

In the event of rosy apple aphid infestation occurring during spring or summer, the shoots should be cut off and removed from the orchard if more than 5% of shoots are affected. Spraying should be carried out after that.

A reduction in winter egg laying by rosy apple aphids can be observed in orchards which lose their leaves quickly because of early termination of shoot growth in autumn.

Rosy leaf-curling aphid (Dysaphis *spp.*)

APPEARANCE.   **Aphid**: grey to dark bluish grey, short black syphon.
Nymphs: pink to bluish grey.
Egg: black.

LIFE CYCLE

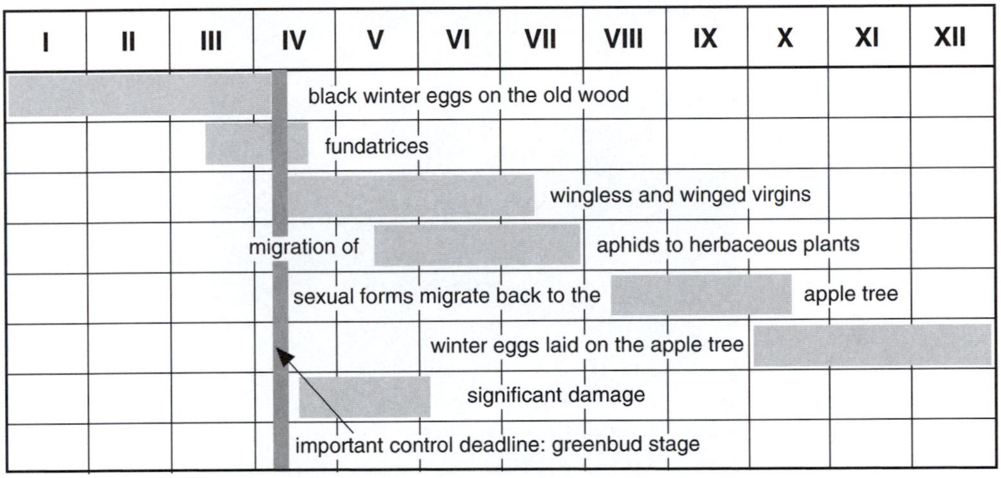

I	II	III	IV	V	VI	VII	VIII	IX	X	XI	XII
| black winter eggs on the old wood											
		| fundatrices									
			wingless and winged virgins								
		migration of	aphids to herbaceous plants								
			sexual forms migrate back to the					apple tree			
			winter eggs laid on the apple tree								
			significant damage								
			important control deadline: greenbud stage								

Fig. 5.28. Life cycle of rosy leaf-curling aphid (*Dysaphis* spp.).

The aphid overwinters as an egg that is laid on the trunk or branches. Soon after the buds open, the larvae hatch, migrate up to the rosette leaves, start to suck and at the same time form their first colonies there. Later the aphids are also found on the young shoots.

They reproduce by parthenogenesis (virgin reproduction). From about mid-May onwards, winged forms are produced which migrate to wild plants. In late summer the females return to the apple tree, where they lay their eggs.

NATURE OF DAMAGE.   First of all there are bright red or yellow spots and galls on the leaves, later folds develop, and the leaves curl up. The fruits show red spots and patches, but these disappear again if there are no aphids left.

ENEMIES.   The rosy leaf-curling aphid has the same enemies as the rosy apple aphid. In the case of the rosy leaf-curling aphid, however, the enemies are much more effective, as the rosy leaf-curling aphid does not have the same reproductive potential.

INSPECTION AND DAMAGE THRESHOLDS.   **First flowering**: 15–20 colonies/100 blossom clusters (blossoms and rosette leaves).

Fig. 5.29. Rosy leaf-curling aphid: damage to a leaf.

Second flowering: 30 colonies/100 shoots (blossoms and leaves).

CONTROL. The rosy leaf-curling aphid is not nearly as dangerous as the rosy apple aphid, but its incidence can increase in hilly locations.

As far as possible, control measures should not be applied immediately, in order to give beneficials (predators and parasites) the possibility of self-regulation. Because the rosy leaf-curling aphid has a large number of enemies, this is easier than in the case of the rosy apple aphid.

If there is no possibility of natural regulation, either a pyrethrum–rotenone formulation or a soap formulation (wetting agent) can be used.

Apple grass aphid (Rhopalosiphum insertum)

APPEARANCE. **Aphid**: bright green body with two longitudinal stripes on the back; the extremities are greyish green.
Eggs: black, laid singly or in groups on old wood.

LIFE CYCLE

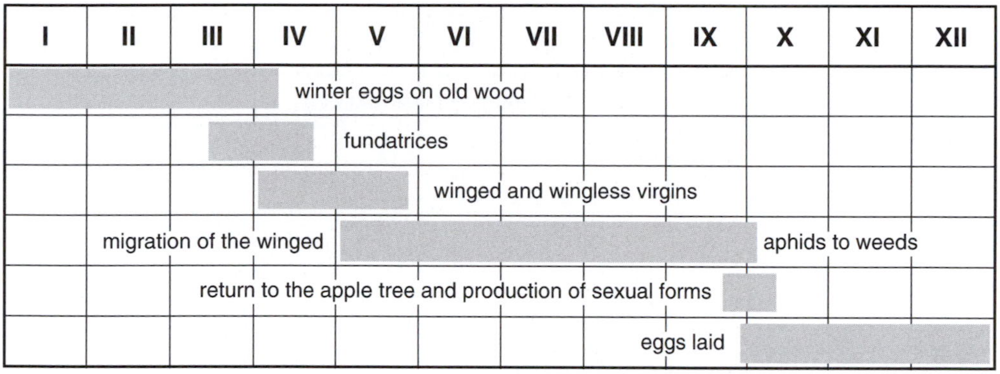

I	II	III	IV	V	VI	VII	VIII	IX	X	XI	XII
winter eggs on old wood											
fundatrices											
winged and wingless virgins											
migration of the winged				aphids to weeds							
return to the apple tree and production of sexual forms											
eggs laid											

Fig. 5.30. Life cycle of apple grass aphid (*Rhopalosiphum insertum*).

The apple grass aphid hatches from the winter eggs as soon as the buds open. The nymphs suck on the rosette leaves or flower buds. At the beginning to middle of May they migrate to weeds (preferring annual meadow grass). The infestation then decreases abruptly. In autumn they return to the apple tree, where they copulate and the females lay the winter eggs on the old wood.

NATURE OF DAMAGE.   There are virtually no visible signs of damage by the apple grass aphid. There may be slight leaf curling, but only if the infestation is very severe. In early spring, however, the aphid may be the first to be observed in large numbers on the flower buds.

ENEMIES.   See section on *Rosy apple aphid.*

DAMAGE THRESHOLD.   The apple grass aphid is an important source of food for all the enemies of aphids and does not actually do any damage.

CONTROL.   Under no circumstances should the apple grass aphid be harmed, as its early occurrence attracts a very large number of beneficials. If large numbers of enemies of aphids appear very early, natural regulation of other aphid species is much easier.

Green apple aphid (Aphis pomi)

APPEARANCE.   **Aphid**: intense green to yellowish green; the abdominal tubes are black.
Egg: black, oblong.

LIFE CYCLE

I	II	III	IV	V	VI	VII	VIII	IX	X	XI	XII
▓	▓	▓	▓	winter eggs in colonies on 1-year-old wood							
			▓	fundatrices							
winged and wingless virgins				▓	▓	▓	▓	▓			
						winter eggs laid		▓	▓	▓	▓
	period of significant damage			▓	▓						

Fig. 5.31. Life cycle of green apple aphid (*Aphis pomi*).

The green apple aphid overwinters as an egg in colonies on the 1-year-old shoot. The fundatrices hatch later than the other aphids. They form their colonies near the apical leaves of the shoot. Under optimal weather conditions there may be massive reproduction from the end of May/beginning of June onwards. At the beginning of June, winged forms are created, which rapidly migrate to other trees. The green apple aphid is the only aphid species that does not change its host. In the autumn, wingless, sexual forms appear, which copulate with each other. The eggs are then laid in large numbers on the bark of the young shoots.

NATURE OF DAMAGE. Severe leaf curling occurs, especially in summer, if there is severe infestation. This is usually accompanied by contamination with sooty mould, which develops on the honeydew secretions of the aphids.

ENEMIES. The same as for the rosy apple aphid.

DAMAGE THRESHOLD. The green apple aphid serves mainly as a food for beneficials and is very rarely harmful. Severe damage can occur only if there is explosive reproduction in nurseries and young orchards.

CONTROL

PREVENTIVE

● Conserve beneficials.
● Avoid excessive shoot production by trees (correct pruning, fertilizer application).

DIRECT. Chemical control of the green apple aphid is seldom necessary if there is optimum conservation of beneficials. If treatment is needed, however, a pyrethrum–rotenone formulation or soap formulation can be used.
Neem has no effect on the green apple aphid.

Fig. 5.32. Green apple aphid on a shoot.

*Woolly aphid (*Eriosoma lanigerum*)*

APPEARANCE. **Aphid**: reddish brown aphids about 2 mm in size, with white, woolly wax secretions; a red juice is excreted when the aphids are squashed.

LIFE CYCLE

I	II	III	IV	V	VI	VII	VIII	IX	X	XI	XII
overwintering nymphs on the trunk and root collar											
			larvae migrate up to pruning wounds, graft unions, etc.								
			8–10 overlapping generations								
			nymphs migrate downwards to the winter retreat								
			significant damage: sticky fruit								

Fig. 5.33. Life cycle of woolly aphid (*Eriosoma lanigerum*).

The naked nymphs, which are not covered with wax, overwinter at the base of the trunk, at the root collar or in cracks in the bark. At the end of March or in April they become active again and migrate up to pruning wounds, graft unions or other wounds and form colonies there under the sticky wax flakes. They produce 8–10 generations asexually. From June onwards, winged forms appear which spread to other trees. In the summer they migrate up to the young shoots and congregate predominantly in the leaf axils. The young shoots can also be attacked by the woolly aphid over their entire length, however.

The rate of reproduction decreases during the summer because of the higher temperatures and the different composition of the sap, but increases again sharply from the end of August onwards.

Wounds (e.g. due to hail) can lead to a very marked increase in woolly aphid infestation.

In Europe, fertile sexual forms cannot develop in the autumn. The semi-adult aphids migrate down again to the winter retreat.

NATURE OF DAMAGE.   Woolly aphid colonies can be recognized by their strong wax secretions. If infestation is severe there are abrasions and lesions on the bark (woolly aphid galls). There is soiling because of the sucking and subsequent honeydew secretion of the aphids.

ENEMIES.   The most important enemy is the wasp-like insect *Aphelinus mali*. This parasitoid has a higher temperature requirement than the woolly aphid and is therefore not as common in the spring. As temperatures rise, the development of the parasitoid proceeds faster, and parasitization levels of up to 90% can be achieved by the autumn. Parasitized aphids do not have the typical wax covering, and the opening where the parasitoid has hatched can be seen in the dead aphids.

Aphelinus mali is sensitive to high levels of sulphur, however.

Other important enemies are the lacewing and the earwig.

DAMAGE THRESHOLD.   The damage threshold is 8–12 colonies per 100 shoots.

CONTROL

PREVENTIVE

- Do not plant susceptible cultivars and rootstocks. Boskoop and Cox's Orange Pippin are considered susceptible, as well as the rootstock M9. With regard to the rootstock, however, for technical reasons there is often no other choice.
- Protect the apple trees from wounds (set up hail nets, etc.).
- Reduce shoot growth: vigorous shoot growth encourages woolly aphids.
- Conserve and encourage beneficials (*Aphelinus mali* and earwigs).

Fig. 5.34. Woolly aphid colony with the typical wax covering.

DIRECT. Chemical control is only necessary in extremely rare instances. With a little patience and a conservative spraying programme, it is possible to achieve a very high degree of parasitization by *Aphelinus mali* in the autumn.

If infestation is too severe, however, efforts can be made to bring it under control with pyrethrum–rotenone formulations or with soap or wetting agents, as severe infestation over a number of years weakens the tree.

San José scale (Quadraspidiotus perniciosus)

APPEARANCE.   **Female**: scale about 1.5 mm, grey.
Male: winged, dark oblong scale.
Larvae: 0.2 mm, yellow, after colony formation the scale is initially white (L1), then black (L2) and finally grey.

LIFE CYCLE.   The larvae in the second larval stage overwinter beneath a small, black scale. Further development into sexually mature forms occurs in the spring. The winged males copulate with the females, which viviparously produce up to 400 yellowish nymphs. In June the young larvae (still without a scale, but with six legs) leave the maternal scale and form colonies on the trunk, usually in the immediate vicinity. After a few days they start to produce their own scale, which is initially white (white-scale stage), then black (black-scale stage) and finally grey. A further generation follows in August. The larvae of the second generation are much more prone to migrate and attack the apples more frequently. These larvae overwinter in the 1st or 2nd stage.

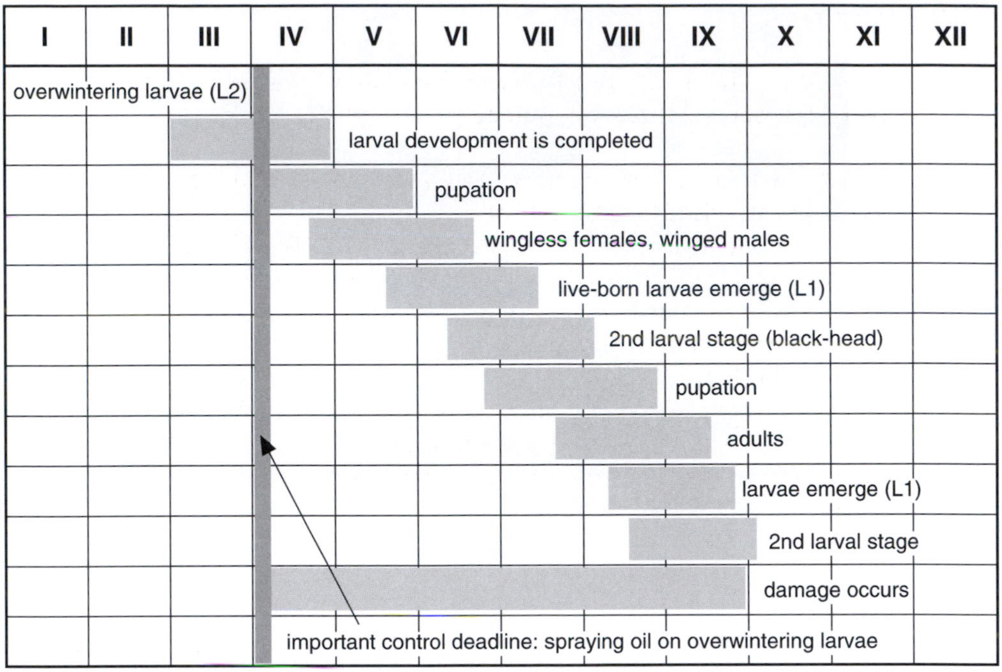

I	II	III	IV	V	VI	VII	VIII	IX	X	XI	XII
overwintering larvae (L2)											
			larval development is completed								
				pupation							
					wingless females, winged males						
						live-born larvae emerge (L1)					
							2nd larval stage (black-head)				
								pupation			
								adults			
									larvae emerge (L1)		
										2nd larval stage	
										damage occurs	
			important control deadline: spraying oil on overwintering larvae								

Fig. 5.35. Life cycle of the San José scale (*Quadraspidiotus perniciosus*).

NATURE OF DAMAGE.   Crusts consisting of grey-coloured scales are found on the parts attacked. If the crust is detached with a knife, the layer of bark or skin beneath is red in colour, because the saliva secreted during sucking reacts with the sap and leads to reddening. This reddening can be very clearly seen, especially on the fruit.

If the attack is very severe, the parts of the plant beneath the crust may die.

ENEMIES.   The most important enemies are the ichneumon wasps of the genera *Encarsia* (*Prospaltella perniciosi*) and *Aphytis*. These small chalcid wasps were released in large numbers in the 1970s and are still commonly found in orchards.

The degree of parasitization by *Prospaltella* can reach 90%.

INSPECTION AND DAMAGE THRESHOLD.   Since the San José scale is a pest subject to quarantine, zero tolerance is applicable, i.e. the damage threshold is exceeded whenever it is present. Inspection at harvesting is of the greatest importance, because the symptoms of damage are easiest to see on the fruit. However, checks should also be made during pruning in winter to see whether the tree is attacked by this pest, so that treatment can still be carried out at budburst if necessary.

CONTROL

PREVENTIVE.   Take care not to harm the chalcid wasps, as they achieve a very high degree of parasitization.

DIRECT.   If there is an attack by San José scale, spraying can be carried out either in winter or at budburst.

Winter spraying: calcium sulphide (200–250 kg/ha) is used for this. Spraying must be carried out during complete winter dormancy (i.e. before the first green leaf appears). At this high concentration the calcium sulphide solution destroys the scale insects. It also controls spider mites and other overwintering stages.

Budburst spraying: paraffin oil is used (30 l/ha). In this operation the scale insects are suffocated by the film of oil.

When to apply: around the mouse-ear/greenbud stage (see also *Control of the red spider mite*).

Codling moth (Cydia pomonella)

APPEARANCE.   **Moth**: 20 mm in size, with grey front wings and darker, wavy transverse lines; there is a wing mark in the apical region – like a dark mirror with three gold-coloured lines in it.
Egg: about 1 mm in size, roundish and very flat, whitish to light green. The egg is laid singly near the fruits on older leaves or directly on the fruits.
Larva: about 15–18 mm long when fully grown, body white with reddish tinge, head reddish brown.

LIFE CYCLE.   The codling moth overwinters as a fully grown caterpillar in a cocoon in cracks in the bark or similar hiding places. The first moths hatch between the middle and end of May. They are active only at dusk, i.e. all activities (flight, feeding, mating, laying eggs) are performed only at dusk. The mated females lay their eggs (80–100 per female) when there are temperatures at dusk of at least 13°C, or preferably 15°C, for several days in succession – first on the leaves and later on the hairless fruits. Warm temperatures and absence of wind encourage this development process, but cold and wet weather can have a very severe effect on egg laying.

The eggs are shiny white at first, then they turn red (red-ring stage); later the black head of the larva can be seen (black-head stage), and finally the caterpillars hatch. This happens about 7–10 days after the eggs are laid.

The young larvae stay for about 1–2 days on the surface of the fruit and then start to burrow into the fruit. They make a small spiral tunnel initially (not as distinct as in the case of *Grapholita lobarzewskii*) and then continue burrowing straight towards the core. The codling moth larva needs to eat the apple pips in order to be able to complete its development. Moist excrement is continuously removed from the tunnel. After more than a month, the development of the larva in the fruit is completed, and the caterpillars leave the apple, usually via the same tunnel, although occasionally an exit tunnel is created. They then migrate down the trunk to cracks in the bark. If the day is still long enough, the caterpillars pupate and a second generation follows in August. If day length is too short, the caterpillars go to their winter refuge.

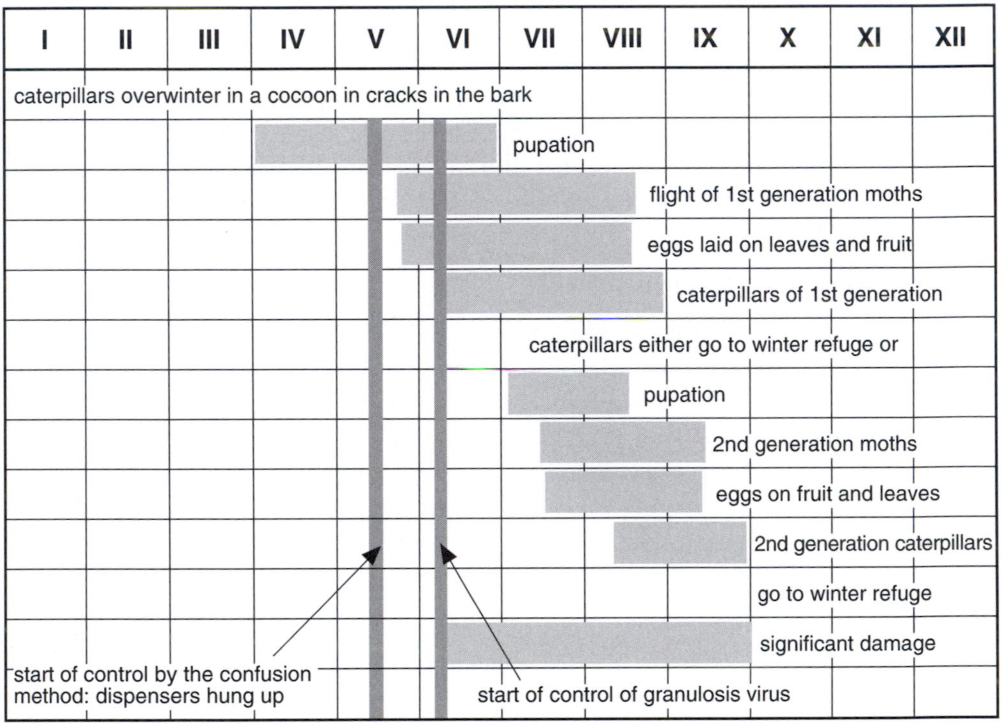

I	II	III	IV	V	VI	VII	VIII	IX	X	XI	XII
caterpillars overwinter in a cocoon in cracks in the bark											
						pupation					
						flight of 1st generation moths					
						eggs laid on leaves and fruit					
						caterpillars of 1st generation					
						caterpillars either go to winter refuge or					
						pupation					
						2nd generation moths					
						eggs on fruit and leaves					
						2nd generation caterpillars					
						go to winter refuge					
						significant damage					
start of control by the confusion method: dispensers hung up						start of control of granulosis virus					

Fig. 5.36. Life cycle of the codling moth (*Cydia pomonella*).

NATURE OF DAMAGE. A tunnel can be seen in the apple which is slightly spiral at the entry point but then leads straight to the core. The edge of the entry aperture turns red. The excrement of the codling moth is moist.

ENEMIES. The codling moth does not actually have any effective enemies, as the caterpillar is very well protected inside the fruit.

The most important enemies of the codling moth are braconid and ichneumon wasps on pupae and fully grown caterpillars. They are not as effective as they are against leaf rollers, however, as the codling moth larva inside the fruit is very well protected against these parasites.

Birds are also less significant here, as the young caterpillars are only on the surface for a short time.

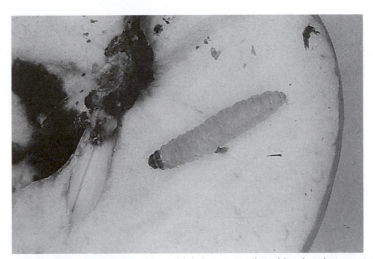

Fig. 5.37. Codling moth caterpillar which has completed its development: it leaves the fruit after eating the pips.

Young codling moth larvae are highly susceptible to infection by viruses and bacteria.

At least as important as the 'live' enemies is another enemy – the weather. The young caterpillars – especially those of the first generation – are highly sensitive to climatic variations. Rain, a drop in temperature or a strong wind after the caterpillars hatch is almost always fatal for the caterpillars. Cool and wet springs or summers can drastically reduce the codling moth risk. The effect of climate is thus much greater than that of natural enemies.

INSPECTION AND DAMAGE THRESHOLD.   The simplest way of monitoring the codling moth is the pheromone trap. The trap needs to be checked every other day. In this way it is possible to determine the start, peak and end of codling moth flight. The numbers caught, however, do not provide any indication of the real magnitude of the damage.

Since in organic fruit growing the control measures are directed either against the moths (confusion method) or the larvae (granulosis virus), the pheromone trap does not yield sufficient information. The hatching date of the larvae cannot be inferred from the flight curve determined by means of the pheromone trap. For this reason it is also essential to observe the hatching of the larvae and/or the burrowing of the larvae into the fruit. This is a very difficult job which needs to be done by skilled professionals, however, so it is best to make use of the information available from the local warning service about the hatching of larvae.

CONTROL.   The codling moth is one of the biggest pests throughout the world, especially in organic orchards. In warm, dry years, in particular, there may be massive reproduction of the codling moth and thus very severe infestation.

For this reason it is important to implement control measures against the codling moth with the utmost care.

PREVENTIVE.   Severe codling moth infestation can often be seen in the vicinity of pallet box storage areas or warehouse compost heaps, if codling-moth-infested fruit have been left in the boxes or dumped together. In this case the fruits should be removed.

DIRECT.   The cleanest method of controlling the codling moth is the **confusion method**.

In this method a cloud of the sexual attractant of the female codling moth is released in the orchard. In this cloud it is impossible for the males to find the female moths and mate with them. Production of progeny is thus prevented.

The following points must be borne in mind, however, if success is to be achieved with this method of codling moth control.

CHOICE OF PLOTS

- At least 3 ha or more
- plots should not be too long or too irregular (square plots are best)
- regular tree volume
- the trees should not be too high.

ISOLATION OF THE ORCHARD

- 100 m from neighbouring sources of infestation with the pest (untreated standard trees)
- buffer zone of 20–40 m in big orchards.

These points are very important because fertilized codling moth females may fly into the orchards using the confusion method, from outside, and cause considerable damage.

INITIAL POPULATION

- Less than 1–2% damage in the past year
- if the initial population is just slightly higher, a treatment must additionally be carried out in the first year.

TIMING OF DISPENSER APPLICATION

- Isomate-C Plus: start of flight
- Ecopom: start of flight and start of July.

DISTRIBUTION OF THE DISPENSERS

- Isomate-C Plus: 1000 dispensers/ha
- Ecopom: 300 dispensers/ha.

The dispensers are mostly placed in the top third of the tree, and closer together at the edge.

In the case of Isomate-C Plus, the number of dispensers can be reduced in the following year, as the active substance is not completely used up.

The confusion method is the best organic approach for control of the codling moth, but in many orchards it cannot be used because the plot is too small. Even if an entire region has been treated with the confusion method, old standard trees remain a potential source of infestation.

All other methods of control are targeted at the larval stage.

USE OF GRANULOSIS VIRUS (CARPOVIRUSINE, MADEX, GRANUPOM, ETC.). The granulosis virus is a highly selective and highly effective formulation for control of codling moth, with a very short action time. The properties of the granulosis virus are as follows.

Granulosis virus formulations have very high infectivity, although the larva feeds very little on the fruit skin. In the laboratory the effect of these formulations (LC50) is 50–100 times greater than that of the best chemical control agents, but in the field they act very slowly. They are highly sensitive to UV light and are very quickly broken down.

The effect can be improved with additives (skim milk powder, pine resin extract (Nu-Film-17)) or by improved formulation.

There may be small, almost invisible scars, as the larvae eat some of the skin before they die.

As the granulosis virus has very high infectivity and a very short life, optimum timing of spraying is of the greatest importance.

The first spraying should be when the larvae hatch. Depending on the weather, further treatments should follow after 5–7 days (if the weather is hot and sunny) or after 8–12 days (if the weather is cool and wet).

The hatching of the larvae can be determined either by visual inspection in the orchard or by using a model to calculate the hatching of the first larvae.

LARVICIDAL EFFECT OF RYANIA (NOT PERMITTED FOR USE IN THE EU AT PRESENT). This alkaloid is also very effective against the larvae of the codling moth. The advantage of Ryania is its broad spectrum of action, allowing control of other caterpillars (tortrix moths, etc.) as well as codling moth.

The disadvantage of Ryania is its high toxicity.

OTHER TORTRIX MOTHS ON FRUIT

Fruitlet mining tortrix *(Pammene rhediella)*

APPEARANCE. **Moth:** about 10 mm; forewings dark brown, reddish brown towards the tip; hind wings dark olive green.

LIFE CYCLE

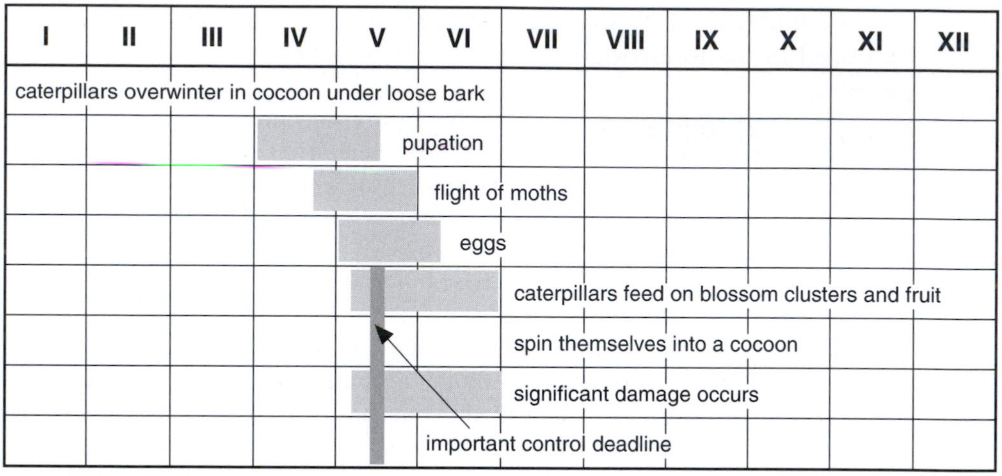

I	II	III	IV	V	VI	VII	VIII	IX	X	XI	XII
caterpillars overwinter in cocoon under loose bark											
					pupation						
					flight of moths						
					eggs						
						caterpillars feed on blossom clusters and fruit					
						spin themselves into a cocoon					
						significant damage occurs					
					important control deadline						

Fig. 5.38. Life cycle of fruitlet mining tortrix moth.

Unlike the codling moth, the moths of the fruitlet mining tortrix are already flying at flowering time. The eggs are laid singly on the underside of leaves or near blossom or fruit clusters. The newly hatched caterpillars often join the stamens and pistils of flowers together with silken webbing. If a single fruit is attacked, the caterpillar often webs a leaf to the surface of the fruit in order to provide protection for itself.

At the end of June the caterpillars spin a cocoon under loose bark, where they overwinter.

NATURE OF DAMAGE. The damage can be seen earlier than that of the codling moth, and also has different characteristics.

The fruitlet mining tortrix often makes more than one hole, with surface damage, and always makes a web. The tunnels do not lead to the core but go in the direction of the base of the stalk, and are more or less free of frass. The fruitlet mining tortrix does not feed on the core.

ENEMIES. The most important beneficial insects are braconid and chalcid wasps.

INSPECTION AND DAMAGE THRESHOLD. Checks at harvesting time are important.

CONTROL

PREVENTIVE. There are no known preventive measures.

DIRECT. Control of the fruitlet mining tortrix is very difficult, as there are no insecticides with a high level of efficacy against this pest.

Ryania is reported to be effective against the young caterpillars, but its use in organic orchards is currently prohibited.

Pyrethrum–rotenone or *Bacillus thuringiensis* sprays have some effect.

Smaller fruit tortrix *(*Grapholita lobarzewski*)*

APPEARANCE. **Moth**: about 12 mm in size, dark brown with a lighter area that is golden or reddish brown in colour.

LIFE CYCLE. The smaller fruit tortrix overwinters as a final larva in the soil. Pupation occurs in the spring and the moths begin to fly around the end of May or beginning of June. The females lay 60–85 eggs about 5 days after flight starts. The young caterpillars make a very distinct spiral path on the fruit and then burrow into the fruit in the direction of the core, but without penetrating into the core. The tunnel usually passes above the core. The caterpillars are active until about mid-August, and the final caterpillar stage moves to the winter refuge from August onwards.

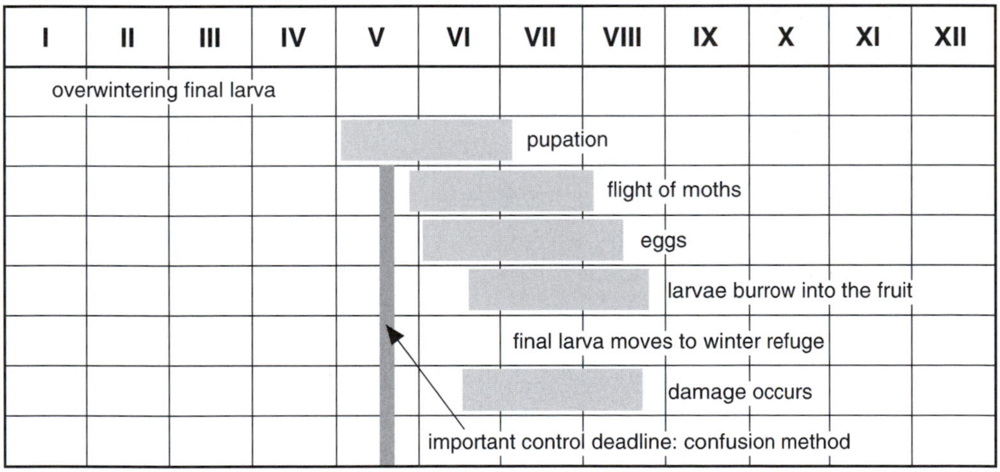

I	II	III	IV	V	VI	VII	VIII	IX	X	XI	XII
overwintering final larva											
				pupation							
					flight of moths						
					eggs						
						larvae burrow into the fruit					
					final larva moves to winter refuge						
						damage occurs					
					important control deadline: confusion method						

Fig. 5.39. Life cycle of smaller fruit tortrix moth.

NATURE OF DAMAGE. The smaller fruit tortrix makes a distinct spiral tunnel at the point of entry into the fruit (in the case of the codling moth the spiral path is much less distinct). The tunnel passes above the core, so the pips are not eaten. The tunnels do not contain frass, and the dry frass hangs threadlike out of the entry hole.

Branching feeding tracks can often be seen in late summer around the point of entry.

ENEMIES.   As with other fruit tortrix moths, chalcid and braconid wasps
are potential enemies. They are not very important, however.

INSPECTION AND DAMAGE THRESHOLD.   It is important to check at harvesting
time the previous year. If infestation exceeds 3–5%, there is likely to be a
major problem with the smaller fruit tortrix the following year.
   The flight of the moths can be monitored with pheromone traps.

CONTROL

PREVENTIVE.   There are no known preventive measures.

DIRECT

- The smaller fruit tortrix is very active on the Gloster cultivar in particular.
- Control with the confusion method can be very effective.

TORTRIX MOTHS THAT FEED ON THE FRUIT

Summer fruit tortrix (Adoxophyes orana = Capua reticulana)

Apple leafroller (Pandemis heperana)

Fruit tree tortrix (Archips podana)

APPEARANCE

Table 5.10. Comparison of summer fruit tortrix, apple leaf-roller and fruit tree tortrix.

Stage of development	Summer fruit tortrix (*Capua reticulana*)	Apple leafroller (*Pandemis heperana*)	Fruit tree tortrix (*Archips podana*)
Moth	7–10 mm, ochre yellow to yellowish brown, irregular brown diagonal bars form a criss-cross pattern with fine brown longitudinal bars	brownish yellow (leather colour) to dark brownish red, with dark oblique band, with a spike pointing towards the base of the wing; a dark, semi-circular patch on the front edge	basic colour brown, the light-brown top third of the wing stands out clearly against the dark-brown, broad oblique band. It is similar to *P. heperana*, but in *Archips* the oblique band is interrupted.
Egg	newly laid eggs are yellow and flattened in shape; they are laid in groups of 80–100 on leaves and sometimes on fruit	newly laid eggs are green, they are laid in clusters on the twig	newly laid eggs are green, they are laid in clusters under a layer of secretion on twigs and branches
Caterpillar	active caterpillar dull green in colour with a honey-yellow head and neck shield	active caterpillar bright green in colour with a green head capsule and green neck shield	squat caterpillar, dark green in colour with a dark brown head capsule and a black neck shield

LIFE CYCLE. These moths have different life cycles. Whereas the summer fruit tortrix (*Capua reticulana*) has two generations, in colder regions of Europe the apple leafroller (*Pandemis heperana*) and fruit tree tortrix (*Archips podana*) often produce only one generation.

DEVELOPMENT OF THE SUMMER FRUIT TORTRIX. The summer fruit tortrix moths overwinter as young larvae in a cocoon on the tree, usually under the remains of dead leaves. Before flowering the first caterpillars move from their winter hiding-places to the flower buds or rosette leaves and web several leaves together in order to feed and grow there. They may also feed on the young fruit. The places where caterpillars have fed can later be seen on the fruit as scars of varying depth.

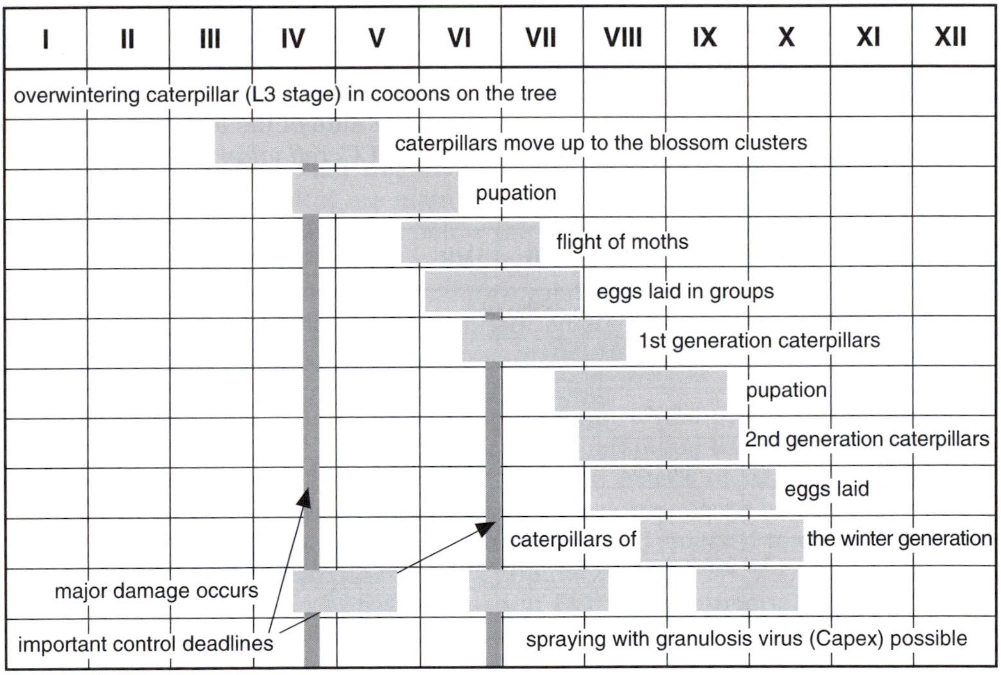

I	II	III	IV	V	VI	VII	VIII	IX	X	XI	XII
overwintering caterpillar (L3 stage) in cocoons on the tree											
			caterpillars move up to the blossom clusters								
				pupation							
					flight of moths						
						eggs laid in groups					
						1st generation caterpillars					
								pupation			
								2nd generation caterpillars			
									eggs laid		
						caterpillars of		the winter generation			
major damage occurs											
important control deadlines					spraying with granulosis virus (Capex) possible						

Fig. 5.40. Life cycle of summer fruit tortrix moth (*Capua reticulana*).

At the end of May the moths of the first generation start to fly. The females lay clusters of 80–100 eggs, covered with a layer of wax, on the leaves. The eggs of the summer fruit tortrix may also be laid on the underside of the leaf.

The caterpillars of the first generation hatch in June. Sometimes they live in the old cocoons of the winter generation, but sometimes they migrate up to the young shoots and web the tips of the shoots together. After 7–10 days they move down to the apples, and the first signs of feeding can be seen on the fruit. The damage caused by the first generation of the summer fruit tortrix can be very severe.

In laboratory and field trials at the Changins Research Institute, Switzerland, P.J. Charmillot found that 10% infestation of shoots corresponds to 1% infestation of the fruit.

SUMMER INSPECTIONS.   In the summer inspections, the main places where the young caterpillars live (gall midge leaves, leaves which have not yet opened at the tips of the shoots, old winter hiding-places, and principal veins on the underside of the leaf) are examined.

The damage threshold is now 10–15% of shoots infested.

CHECKS FOR INFESTATION OF FRUIT IN SUMMER.   It is important to know the populations on the fruit and on the leaves, as some caterpillars do not feed on the fruit but may complete their development on the leaves.

About 500–1000 apples should be checked. The damage threshold is 2% of apples infested. It is very dangerous if this threshold is exceeded, as the caterpillars usually live in old cocoons, where they are difficult to control with plant protection products.

CONTROL.   The tortrix moths that feed on fruit have become less of a problem in integrated and especially in organic fruit production. Very high rates of parasitism by chalcid and braconid wasps can be achieved by conservation of beneficial insects.

PREVENTIVE

* Creation of nesting boxes for tits.
* Conservation of chalcid and braconid wasps.
* Early cessation of shoot growth (tortrix moths attack the tip of the shoot).

DIRECT.   If there is severe infestation in autumn (>3% of fruits attacked), spraying with granulosis virus (Capex) should be carried out the following spring in the greenbud–redbud stage. Capex is very effective against the summer fruit tortrix. Care should be taken to make sure that there is suitable UV stabilization of the spray.

Capex can also be used against the summer generation. Capex is more effective against the young caterpillars, so spraying should be carried out soon after hatching.

Pyrethrum–rotenone sprays also have some effect on other tortrix moths.

NOCTUID MOTHS

Orthosia *spp*.

Cosmia trapezina

Amphipyra pyramidea

APPEARANCE.   **Moths:** 30–40 mm in size, usually greyish brown to reddish brown, variable in colour and pattern.

Eggs: usually laid in groups of 20–100 on fruit-bearing branches.
Larvae: the larvae of the noctuid moths have the same number of thoracic legs and abdominal feet as the tortrix caterpillars, but their head is rounded and the mouth parts point downwards. When the caterpillar is disturbed it curls up in a C shape.

LIFE CYCLE

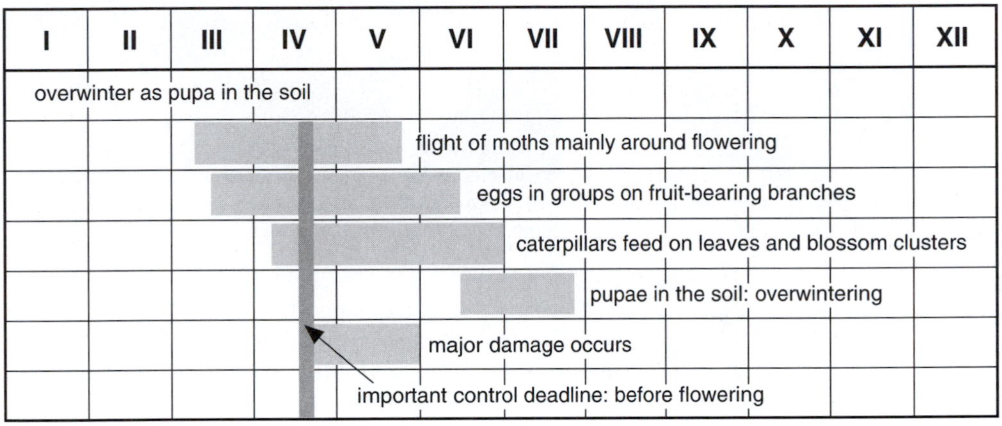

I	II	III	IV	V	VI	VII	VIII	IX	X	XI	XII
overwinter as pupa in the soil											
					flight of moths mainly around flowering						
					eggs in groups on fruit-bearing branches						
					caterpillars feed on leaves and blossom clusters						
					pupae in the soil: overwintering						
					major damage occurs						
					important control deadline: before flowering						

Fig. 5.44. Life cycle of noctuid moths.

Noctuid moths overwinter as pupae in the soil. The moths fly before and during flowering at dusk and lay their eggs on the fruit-bearing branches. The young caterpillars feed on the leaves, making the typical holes and damage at the edge of the leaves, and they also damage the fruit. Unlike tortrix caterpillars, they do not live in a cocoon, but it is very difficult to find them, since they feed only at night. Only the youngest caterpillars sometimes make a very small cocoon at the tips of the shoots – older caterpillars never do this. The caterpillars reach adulthood between the middle and end of June and then move down to the soil, where they pupate.

Noctuid moths are polyphagous, i.e. they also develop in many different types of broad-leaved trees (oak, lime, poplar, etc.), from which they can fly into the orchard in spring.

NATURE OF DAMAGE. The typical holes or damage at the edge can be seen in the leaves. Blossom clusters are also attacked, but no cocoon is found there.

Very deep damage, which becomes corky by autumn, is often found on fruit, as in the case of the winter moth. At harvesting time it is no longer possible to distinguish the damage from that caused by other caterpillars.

ENEMIES. Enemies include tits, tachinid flies and braconid wasps. Tits, in particular, can be very effective enemies in the spring.

Fig. 5.45. Apple at harvesting time with characteristic noctuid damage.

Fig. 5.46. Winter moth caterpillar moving with typical looping action.

INSPECTION AND DAMAGE THRESHOLD.   The damage threshold is not usually reached before flowering. Control measures need to be applied before flowering, however, because the damage to the fruit is already caused before or during flowering; this means that the extent of the infestation at harvesting time is used as the basis for assessment of the damage threshold for pre-flowering treatment.

If more than 2% of the fruit is infested at harvesting time, pre-flowering treatment in the following year is advisable (as demonstrated in a study by P.J. Charmillot at Changins, Switzerland).

After flowering, the damage threshold is 5% of blossom clusters infested.

The flight of the moths can be monitored with pheromone traps, but a specific attractant has to be used for each species.

CONTROL.   Noctuid moths often occur in concentrations around a focal point, and severe damage may be caused within such a concentration. Orchards near to woodland and forests are especially at risk.

PREVENTIVE.   No preventive measures are known.

DIRECT.   Pyrethrum–rotenone sprays have some effect on noctuid moths. At high temperatures *Bacillus thuringiensis* formulations also have some effect on noctuid moths (50% effect).

Winter moth *(Operophtera brumata)*

APPEARANCE.   **Female moth**: the wings are reduced to a stump; the body is 5–6 mm long and has a mottled dark-brown to yellowish grey colour.
Male moth: wing span 22–28 mm, forewings greyish brown with wavy cross-lines.

Egg: 0.4–0.5 mm long, oval, yellowish green at first, turning to orange later. **Caterpillar**: up to 25 mm long, coarse, bright green with a dark green line on the back. The caterpillar only has two abdominal feet and moves with a looping action.

LIFE CYCLE

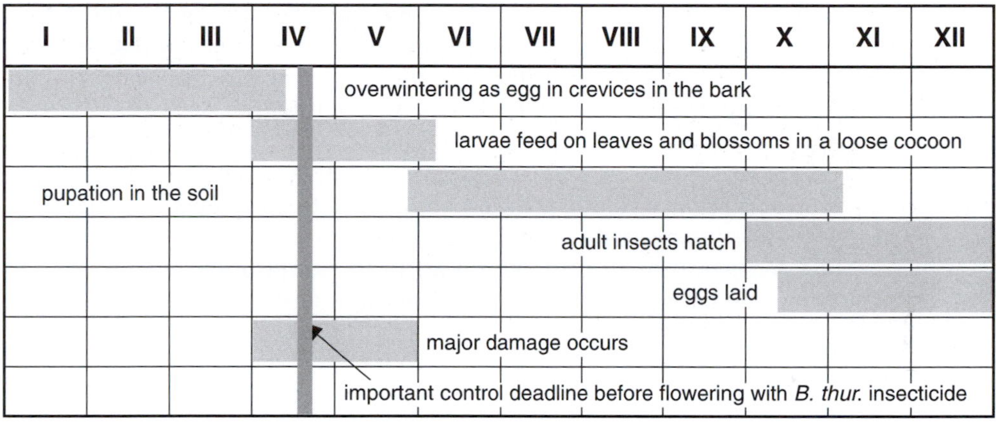

I	II	III	IV	V	VI	VII	VIII	IX	X	XI	XII
\multicolumn{12}{} overwintering as egg in crevices in the bark											
\multicolumn{12}{} larvae feed on leaves and blossoms in a loose cocoon											
\multicolumn{12}{} pupation in the soil											
\multicolumn{12}{} adult insects hatch											
\multicolumn{12}{} eggs laid											
\multicolumn{12}{} major damage occurs											
\multicolumn{12}{} important control deadline before flowering with *B. thur.* insecticide											

Fig. 5.47. Life cycle of the winter moth (*Operophtera brumata*).

The moths hatch from October onwards. The flight of the males begins immediately, whereas the females are unable to fly. After hatching, the female moths move up the trunk and lay up to 300 eggs in crevices in the bark. The young caterpillars hatch from bud break until flowering. They have three pairs of thoracic legs but only two pairs of abdominal legs, and begin to feed immediately on the young rosette leaves and flower buds. They web the young leaves loosely together and live in them. Infested blossoms are eaten almost completely. Older caterpillar stages also feed on the fruit. This damage is not as deep as in the case of noctuid moths, however.

When the caterpillars are fully grown they leave the tree and pupate in the soil in a cocoon.

NATURE OF DAMAGE. On the leaves and fruit there is damage due to feeding, with a small cocoon. A spoon-like cavity may be eaten out of the fruit. Older caterpillars skeletonize the leaves and make holes in them.

ENEMIES. Birds, especially coal-tits, are important enemies of winter moth caterpillars. The feeding requirements of these birds at brood time are very high (300 caterpillars/day for the brood). Coal-tits are capable of keeping winter moth infestation below the damage threshold.

Parasitization of the caterpillars by ichneumon wasps is also very important.

INSPECTION AND DAMAGE THRESHOLD.   Winter moth caterpillars are detected by visual inspection, with 100 blossom clusters being examined. The damage threshold is 10–15% of blossom clusters infested.

CONTROL

PREVENTIVE.   Establishment of nesting boxes for tits (especially coal-tits).

Because of the labour costs involved, in intensive orchards with a high tree density it is now almost impossible to place grease bands around the trunk to prevent the females crawling up the trunk in the autumn. In old orchards or in orchards with a low tree density, however, this may be a useful aid.

DIRECT.   Under optimum conditions *Bacillus thuringiensis* insecticides are very effective against winter moth caterpillars.

The temperature should be 15–20°C for several days, however, as otherwise the caterpillars do not feed enough and do not ingest enough of the active substance.

BUD MOTHS

Apple bud moth *(*Spilonota ocellana*)*

Green bud moth *(*Hedya nubiferana*)*

APPEARANCE

Table 5.11. Comparison of apple bud moth (*Spilonota ocellana*) and green bud moth (*Hedya nubiferana*).

Stage of development	*Spilonota ocellana*	*Hedya nubiferana*
Moth	14–18 mm, fore and rear wings brown, the forewings are whitish in the middle	17–21 mm, forewings dark brown and bluish grey, the tip is whitish with a dark pattern, rear wings greyish brown
Caterpillar	9–12 mm long, dark purple brown with lighter nodules	18–20 mm long, olive green to dark green with black nodules

LIFE CYCLE.   They overwinter as semi-adult caterpillars in crevices in the bark or beneath bud scales. They emerge at budburst and feed on the rosette leaves and blossom trusses, which they often web together. A dried leaf is usually incorporated in this web. The caterpillar of *Spilonota ocellana* develops rather more slowly than that of *Hedya nubiferana* and occasionally feeds on the young fruit beneath a leaf that is webbed to it. These wounds form cork and can later be seen as malformations of the fruit. The moths fly in June and July and lay their eggs singly on the leaves. The caterpillars which hatch out of the eggs live until autumn on leaves webbed together and can also attack fruit.

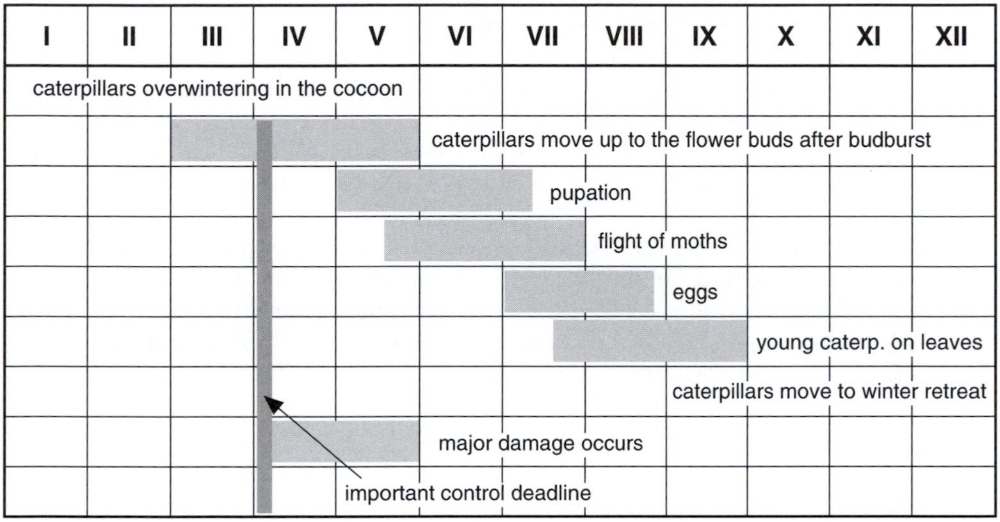

I	II	III	IV	V	VI	VII	VIII	IX	X	XI	XII
caterpillars overwintering in the cocoon											
					caterpillars move up to the flower buds after budburst						
					pupation						
						flight of moths					
						eggs					
							young caterp. on leaves				
							caterpillars move to winter retreat				
					major damage occurs						
					important control deadline						

Fig. 5.48. Life cycle of bud moths.

NATURE OF DAMAGE.   The damage can already be seen at an early stage. Hollowed-out buds are found, usually together with a web.

ENEMIES.   Braconid and ichneumon wasps, and birds.

INSPECTION AND DAMAGE THRESHOLD.   Usually an opportunistic pest. Inspection must take place before flowering, as after flowering it is too late.

The damage threshold is 5–8% of blossom clusters infested in the case of *Spilonota ocellana* and 10–15% of blossom clusters infested in the case of *Hedya nubiferana*.

CONTROL

PREVENTIVE.   There are no known preventive measures.

DIRECT.   Control measures must be applied before flowering, after the damage threshold is exceeded. *Bacillus thuringiensis* and pyrethrum–rotenone formulations only have a limited effect in organic orchards, however.

LEAF-MINING MOTHS

Pear leaf blister moth (Leucoptera scitella)

Spotted tentiform leaf-miner (Phyllonorycter blancardella)

Peach leaf-miner (Lyonetia clerkella)

Apple leaf-miner (Nepticula malella)

Hazelnut leaf-miner (Lithocolletis corylifoliella)

APPEARANCE AND DAMAGE CAUSED

Of the leaf-mining moths, only *Leucoptera scitella* is really dangerous. This moth will therefore be considered in greater detail.

LIFE CYCLE. *Leucoptera scitella* overwinters as a pupa in a whitish cocoon. Flight of the moths begins around flowering. The females lay the eggs on the underside of the leaves, especially near the trunk. The conspicuously segmented caterpillar penetrates into the leaf and lives between the wax layer and epidermis. After about 3 weeks the first mines, which slowly get bigger, can be seen on the leaf surface. Pupation of the caterpillars takes place in June near the mines. The flight of the second generation starts at the end of June/beginning of July. Infestation by the second generation is much more severe, and a third generation may be produced if weather conditions are good.

ENEMIES. The most important enemies of the leaf-mining moths are braconid and chalcid wasps. These are able to keep all the leaf-mining moth species except *Leucoptera scitella* below the commercial damage threshold.

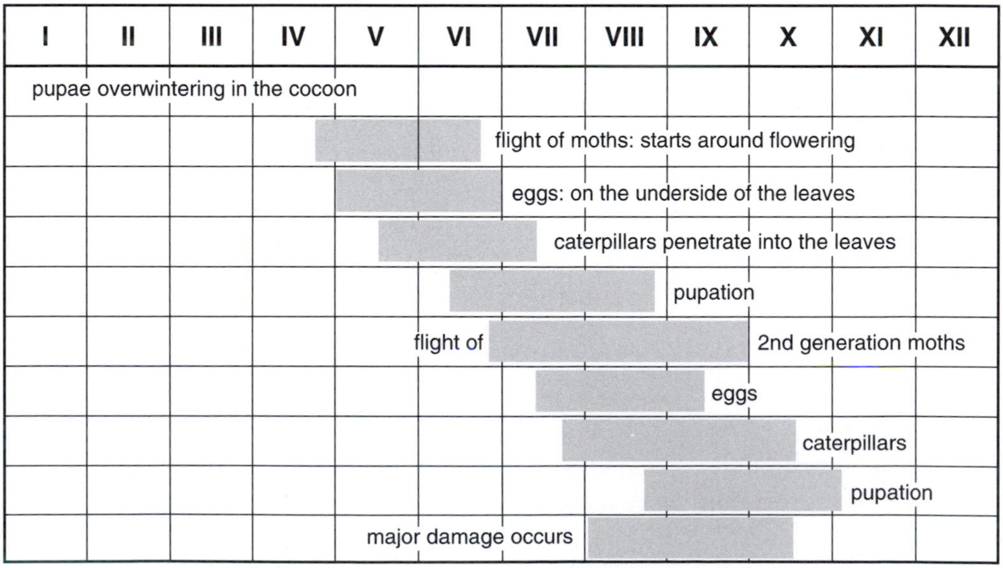

I	II	III	IV	V	VI	VII	VIII	IX	X	XI	XII
pupae overwintering in the cocoon											
				flight of moths: starts around flowering							
				eggs: on the underside of the leaves							
				caterpillars penetrate into the leaves							
					pupation						
				flight of 2nd generation moths							
						eggs					
						caterpillars					
						pupation					
				major damage occurs							

Fig. 5.49. Life cycle of pear leaf blister moth (*Leucoptera scitella*).

Table 5.12. Comparison of leaf-mining moths: pear leaf blister moth (*Leucoptera scitella*), spotted tentiform leaf-miner (*Phyllonorycter blancardella*), peach leaf-miner (*Lyonetia clerkella*), apple leaf-miner (*Nepticula malella*) and hazelnut leaf-miner (*Lithocolletis corylifoliella*).

	Leucoptera scitella	*Phyllonorycter blancardella*	*Lyonetia clerkella*	*Nepticula malella*	*Lithocolletis corylifoliella*
Appearance of moths	3–4 mm, silvery, with black-and-yellow pattern on the forewing	3–4 mm, copper brown wings with white pattern with dark margins	3–4 mm, white, mottled brown forewings	2 mm, dark, there is a brilliant white cross-band in the rear part of the wing	4 mm, chestnut brown, partially interspersed with black
Generations	2–3	2–4	3	3	2–3
Overwintering	pupa in whitish cocoon	pupa in the mines	moth in natural refuges	pupa in brown cocoon (fallen leaves)	fully grown larva in the mines
Nature of damage	roundish, coin-like mines on the top of the leaf	light patch with lengthwise fold on underside of leaf, criss-cross pattern on top of leaf	serpentine mines, about 5 cm long, which gradually become wider	narrow mines which quickly become wider, short, reddish brown frass trail can be seen through the leaf surface	pale patch 1–2 cm in size with small rust patches on the top of the leaf
Level of danger 1st moths	very dangerous end of April–start of May	not very dangerous April	not very dangerous end of April	not very dangerous end of March–May	not very dangerous April–May
Time when damage occurs	May–September	May–September	May–September	May–September	May to mid-June, mid-July to September

Parasitization levels of 80–90% can be achieved by autumn. Only if weather conditions are poor in the spring (cold, wet weather) is the development of the parasites retarded.

INSPECTION AND DAMAGE THRESHOLD.   Occurrence of the pest is monitored by visual inspection of the leaves.

Inspection at harvesting time is particularly important, but inspections should also be made in spring and summer.

If the level of infestation by *Leucoptera scitella* is 3–5 mines/tree in the first generation, there is likely to be a big increase in infestation in the second generation.

The flight of the moths can be monitored by means of pheromone traps. In this way it is possible to obtain an overall picture of the presence and species composition of the leaf-mining moths.

CONTROL

PREVENTIVE.   Conservation of beneficial insects (chalcid and braconid wasps).

DIRECT.   Direct control is impossible in organic orchards, as no control agents are available. In any case it is only necessary in very rare cases, as infestation is usually kept below the damage threshold by taking advantage of natural limiting factors (parasites, abrupt temperature variations, cold winters).

CATERPILLARS LIVING ON WOOD AND BARK

Leopard moth (Zeuzera pyrina)

Goat moth (Cossus cossus)

APPEARANCE AND DAMAGE CAUSED

Table 5.13. Comparison of leopard moth (*Zeuzera pyrina*) and goat moth (*Cossus cossus*).

Stage of development	Leopard moth (*Zeuzera pyrina*)	Goat moth (*Cossus cossus*)
Moth	about 4.5–6.5 cm in size; white wings with bluish-black spots	about 7–10 cm in size, greyish brown with irregular black pattern
Caterpillar	5–6 cm long, first pink, later yellowish white with small black spots	8–10 cm long, first pink, later the sides are yellowish and the back is dark red; the head is black
Flight of moths	at night, June/July	at night, June to beginning of August
Egg laying	singly on twigs of younger branches, especially on leaf stalks and buds	in clusters of up to 20, in crevices in bark or in lesions in the trunk

Continued

Table 5.13. *Continued.*

Stage of development	Leopard moth (*Zeuzera pyrina*)	Goat moth (*Cossus cossus*)
Eggs produced per female	about 800	> 1000
Caterpillar development	caterpillars live first in the pith of twigs, eat out a space under the bark of stronger wood and then tunnel upwards in the pith	initially the caterpillars feed together in a space between the bark and wood, then in the next spring (April) each caterpillar makes its own feeding tunnel in the wood, eventually over 1 m in length
Nature of damage	leaves or shoots wither and die, frass pellets and wood particles protrude from the entry hole	the feeding tunnels are extensive and significantly larger than those of the leopard moth, the infestation is normally near the trunk. The caterpillar gives off an unpleasant odour

LIFE CYCLE. Both moths have a development cycle that lasts several years.

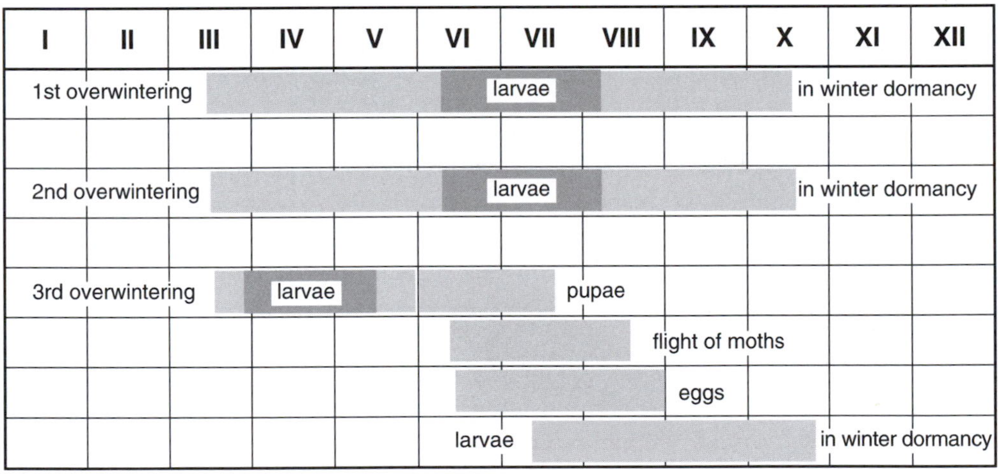

Fig. 5.50. Life cycle of leopard moth (*Zeuzera pyrina*) and goat moth (*Cossus cossus*).

ENEMIES. No significant enemies are known.

INSPECTION AND DAMAGE THRESHOLD. There is no damage threshold for these pests, but even just a few caterpillars can cause a great deal of damage.

Visual inspection of the trees for traces of tunnelling dust.

The flight of the moths can be monitored with pheromone traps. However, these only give an approximate idea of the presence of moths in the orchard.

CONTROL

PREVENTIVE. Avoid damaging the trunk (with hoeing tools and mulchers).

DIRECT. The only direct control measure is insertion of a wire into the entry hole for mechanical destruction of the caterpillars. For control of the leopard moth, however, positive results have also been obtained with mass traps (12 pheromone traps/ha).

Apple clearwing moth *(Synanthedon myopaeformis)*

APPEARANCE. **Moth**: 20–25 mm in size, transparent wings, black body with broad red band across it.
Caterpillar: 15–17 mm long, creamy white.

LIFE CYCLE. The apple clearwing moth has a development cycle that extends over several years. The eggs are laid in June/July in crevices in the bark on the trunk or at the base of the main branches, often near wounds or old sites of infestation. As soon as they hatch, the caterpillars bore into the twigs and make irregular meandering tunnels in or just below the bark. As the caterpillars take 20 months to develop, old and young caterpillars are often found on the same tree. When their development is completed, they spin tough grey cocoons beneath the surface of the bark between decomposing wood and caterpillar frass.

NATURE OF DAMAGE. Irregular tunnels at the base of the trunk; the bark peels off more easily at the sites of infestation. If there is severe infestation, sticky sap oozes out of the bark and the sites of infestation show blackish spots.

ENEMIES. The main enemies are birds.

INSPECTION AND DAMAGE THRESHOLD
- Use of pheromone traps to assess the degree of infestation.
- Visual inspections for entrance holes around the trunk.

CONTROL

PREVENTIVE. Avoidance of injuries.

DIRECT. The apple clearwing moth can be controlled very effectively with the confusion method.

Apple blossom weevil *(Anthonomus pomorum)*

APPEARANCE. **Beetle**: 3.5–6 mm long, dark brown to black; on the wing covers there is a white mark in the shape of a V, open towards the front.
Egg: 0.5–0.7 mm long, oval, white and translucent.

Fig. 5.51. Leopard moth caterpillar inside a branch.

Fig. 5.52. Apple clearwing moth: a characteristic feature is the orange band across the body.

Larva: 8 mm long, white with a dark brown head. The body of older larvae is yellow.

LIFE CYCLE.   The apple blossom weevil overwinters in crevices in bark, in piles of wood, under leaves or in other dry, sheltered places outside the orchard, especially in woodland.

I	II	III	IV	V	VI	VII	VIII	IX	X	XI	XII
winter dormancy				first beetles appear on the first warm days in March							
				eggs laid mainly in the mouse-ear stage							
				larva: protected in the capped blossom							
					pupation in the blossom						
						beetles feed and go to the winter hiding-place					
						damage caused by the beetles					
			important control deadline: beetles before the eggs are laid								

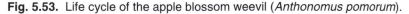

Fig. 5.53. Life cycle of the apple blossom weevil (*Anthonomus pomorum*).

With the first warm, sunny days in spring (end of February/beginning of March) the beetle appears and begins to bore into the blossom buds that are opening. The apple blossom weevil mainly attacks early- and medium-early-flowering apple varieties (e.g. Idared, McIntosh, Jonagold). Its rostrum is a highly sensitive probing instrument, and it uses it to try to find the ideal development stage for egg-laying. In

severe attacks the holes bored in the blossom buds can later be found
in the apples as indentations near the calyx. The eggs are laid in the
buds at about the mouse-ear stage. The beetle only lays one egg in
each blossom. After 7–10 days the grubs hatch and immediately eat
the style and anthers of the blossom. To prevent the blossom from
opening and make sure that the grub in the blossom is sheltered from
the sun, the petals are eaten away at the base. This gives rise to the
characteristic damage caused, in the form of brown, balloon-like blossoms which fail to open.

After about 1 month, the grubs are fully grown and pupate in the
blossom. About 2–3 weeks later, the young beetles hatch. After a short
period to allow their shells to harden they leave the blossoms and begin
to feed on the leaves and young fruit.

After 2–3 weeks the beetles move to their winter hiding-places.

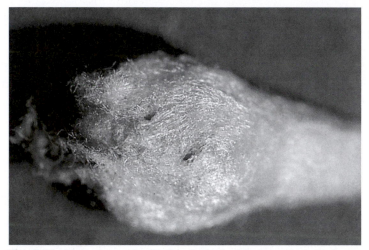

Fig. 5.54. Holes bored by the beetle in the blossom buds that are opening.

NATURE OF DAMAGE.   The holes bored by the beetles are visible in the blossom buds at budburst.

If an egg has been laid in the blossom, the typical brown, balloon-
like blossoms which fail to open can later be seen.

Two types of damage can subsequently be seen in the fruit: firstly the
indentations near the calyx, due to the holes that were bored at an early
stage, and secondly new holes bored by the young beetles both in the
fruit and in the leaves. The latter type of damage is conspicuous only if
there is a very high population density.

ENEMIES. The apple blossom weevil does not have any enemies that control it really effectively. The egg, grub and pupa are very well protected in the blossom.

Enemies which are frequently mentioned are fungi, chalcid wasps and nematodes, which attack and parasitize the beetles in their winter hiding-places.

Birds, which eat the beetles in spring, only have a limited effect.

INSPECTION AND DAMAGE THRESHOLD. The apple blossom weevil is monitored mainly by means of the knockdown test.

This test must be carried out in warm weather with little wind (preferably at midday, when the temperature is above 10°C), as otherwise the result does not provide reliable information.

The damage threshold is 10–40 beetles/100 branches tested (10 beetles if flowering is poor, 40 beetles if flowering is good).

The widely held opinion that the apple blossom weevil is a welcome blossom thinner is not applicable to modern organic fruit production. The blossom weevil cannot be controlled and can thus cause damage even when there is abundant production of blossom buds.

The laying of the eggs can be monitored by visual inspection. This is done by collecting the individual buds which have been pierced and carefully opening them with a needle.

CONTROL

PREVENTIVE. No special measures are known.

DIRECT. After the damage threshold is exceeded, a pyrethrum–rotenone spray should be applied at temperatures above 12°C. If very early treatment is necessary because of the high population density, a second spraying may sometimes be needed, as beetles move continuously from the winter hiding-places to the apple trees until the eggs are laid.

The knockdown tests should therefore be continued after spraying.

OTHER WEEVILS THAT ATTACK POME FRUIT

Brown leaf weevil (Phyllobius oblongus)

Beech leaf-mining weevil (Orchestes fagi = Rhynchaenus fagi) and other Phyllobius species

APPEARANCE. Weevils that are 3–6 mm in size, of various colours (the brown leaf weevil is brown with a black head and neck-shield, the beech leaf-mining weevil is black or dark brown, and the green weevil is green).

LIFE CYCLE

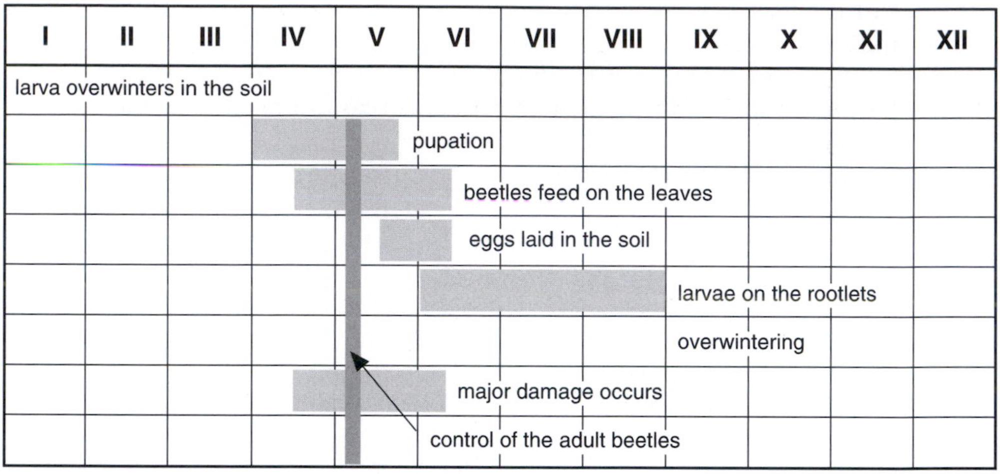

I	II	III	IV	V	VI	VII	VIII	IX	X	XI	XII
larva overwinters in the soil											
					pupation						
					beetles feed on the leaves						
					eggs laid in the soil						
								larvae on the rootlets			
								overwintering			
					major damage occurs						
					control of the adult beetles						

Fig. 5.55. Life cycle of the brown leaf weevil (*Phyllobius oblongus*).

These beetles overwinter as larvae in the soil. Around apple blossom time the beetles emerge from their winter hiding-places and feed on the leaves and petals. The leaves are eaten from the margin inwards. These beetles can cause considerable damage, especially in nurseries and young orchards.

The beetles lay their eggs in the soil from the end of May onwards. The young larvae feed on the rootlets, but without causing any real harm.

The beech leaf-mining weevil has a rather different life cycle. In this case the weevil overwinters as a beetle. The eggs of the beech leaf-mining weevil are laid on the leaves of beech trees. It is only occasionally found in orchards.

NATURE OF DAMAGE.   Holes and skeletonization of leaves. The beech leaf-mining weevil may also make shallow, roundish holes in fruit.

ENEMIES.   The principal known enemies are birds.

INSPECTION AND DAMAGE THRESHOLD.   Infestation is monitored by means of the knockdown test. The brown leaf weevil has occurred in very large numbers in recent years. Knockdown tests with 200–300 brown leaf weevil beetles/100 branches are quite common.

The damage threshold is 100–150 beetles before flowering and 200–250 beetles after flowering. Young orchards and nurseries should be monitored with particular care.

In the case of the beech leaf-mining weevil, it is particularly important to monitor orchards near woodland.

CONTROL

PREVENTIVE.   There are no known preventive control measures.

DIRECT.   Application of pyrethrum–rotenone sprays after the damage threshold is exceeded.

Apple fruit rhynchites (Rhynchites aequatus)

APPEARANCE.   Beetle that is 3–5 mm in size, with red or reddish brown forewings and a stumpy shape.

LIFE CYCLE

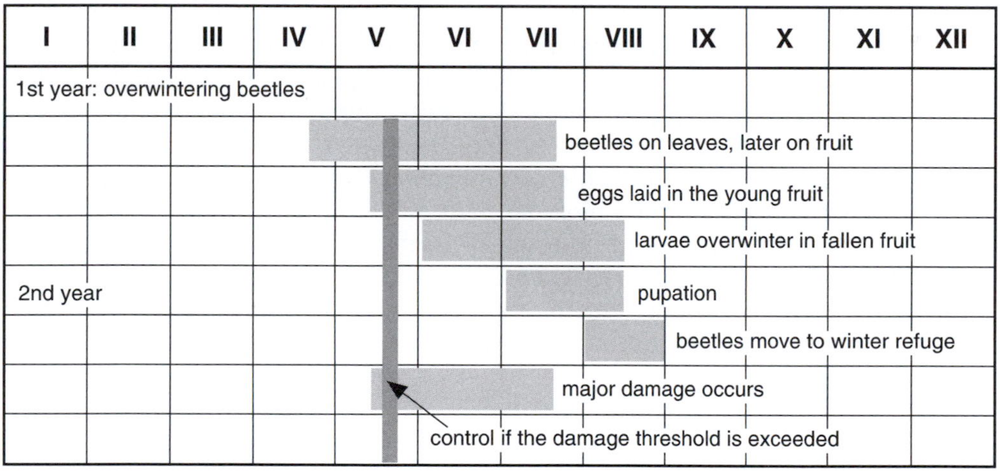

I	II	III	IV	V	VI	VII	VIII	IX	X	XI	XII
1st year: overwintering beetles											
						beetles on leaves, later on fruit					
						eggs laid in the young fruit					
							larvae overwinter in fallen fruit				
2nd year						pupation					
							beetles move to winter refuge				
						major damage occurs					
					control if the damage threshold is exceeded						

Fig. 5.56. Life cycle of apple fruit rhynchites (*Rhynchites aequatus*).

The apple fruit rhynchites has a 2-year life cycle. The beetles appear in May and feed initially on the leaves or buds and later preferentially on the young fruit, making large numbers of holes in the flesh. The eggs are laid in the fruit from June onwards (single entry holes). The larvae eat their way into the core, and the wrinkled fruit falls in autumn or earlier. The larvae overwinter in this fallen fruit and pupate in the soil in the following June. The young beetles immediately move to their winter hiding-places in the soil.

NATURE OF DAMAGE.   From May onwards, damage due to feeding can be seen at a number of places in the fruit, or there are single entry holes (for laying of eggs) which are later covered by scar tissue.

ENEMIES. Birds.

INSPECTION AND DAMAGE THRESHOLD. The damage threshold is determined by visual inspection. It is about 2–3 beetles/100 fruit.

CONTROL. The pyrethrum–rotenone sprays used against the apple blossom weevil are also effective against the apple fruit rhynchites. The control deadline is much later, however (May–June).

Pear bud weevil *(Anthonomus piri)*

APPEARANCE. **Beetle**: 4.5–6 mm long, brownish with a light 'cross-band'. **Larva**: 5–7 mm long, creamy white with dark-brown head, more pronounced C shape than the larva of the apple blossom weevil.

LIFE CYCLE. The beetles hatch in May and feed on the tips of shoots and on leaf stalks. They soon look for a hiding-place, however, and are dormant throughout the summer. Not until the start of September do the beetles re-emerge from their hiding-places and start to eat out cavities at the base of the buds, which consequently die. The mated females lay their eggs singly in the blossom buds, which they have already pierced. Oviposition may continue until December and in many years until the following spring. The larvae hatch at the end of October (or not until the following spring), eat out the interior of the blossom bud and pupate at the end of April/beginning of May.

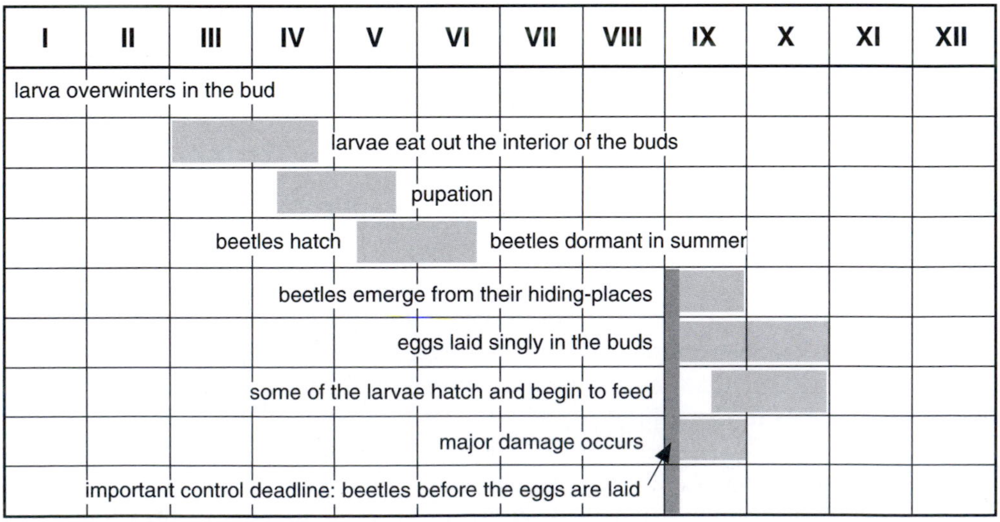

I	II	III	IV	V	VI	VII	VIII	IX	X	XI	XII
larva overwinters in the bud											
		▓	▓ larvae eat out the interior of the buds								
			▓	▓ pupation							
		beetles hatch		▓	▓ beetles dormant in summer						
			beetles emerge from their hiding-places					▓			
				eggs laid singly in the buds				▓	▓		
			some of the larvae hatch and begin to feed						▓	▓	
					major damage occurs			▓			
	important control deadline: beetles before the eggs are laid										

Fig. 5.57. Life cycle of pear bud weevil (*Anthonomus piri*).

NATURE OF DAMAGE.   The leaf and blossom buds do not open in the spring. These buds are still closed while the other buds are already coming into flower. The bud scales are not completely closed but rather lacerated.

ENEMIES.   No major enemies are known.

INSPECTION AND DAMAGE THRESHOLD.   Monitoring is very difficult, as the beetles cling to the branches very securely and are very difficult to detect with the knockdown test. The infestation in the previous year is very important. Infestation can increase very severely in 2–3 years and, starting from just a few trees, spread to the entire orchard.

CONTROL

PREVENTIVE.   No preventive control measures are known.

DIRECT.   If there has been infestation in the previous year, the first spraying must be done in September, immediately after the harvest, using pyrethrum–rotenone sprays. The spraying must be repeated if necessary. It is important that the weather is warm and fine.
    It is too late for control measures once the eggs are laid.

European shot hole borer *(Xyleborus dispar)*

APPEARANCE.   **Females:** 3.0–3.5 mm long, winged, dark brown to black, yellowish hairs on body.
Males: 1.5–2.0 mm long, incapable of flight, short abdomen and small thorax.

LIFE CYCLE

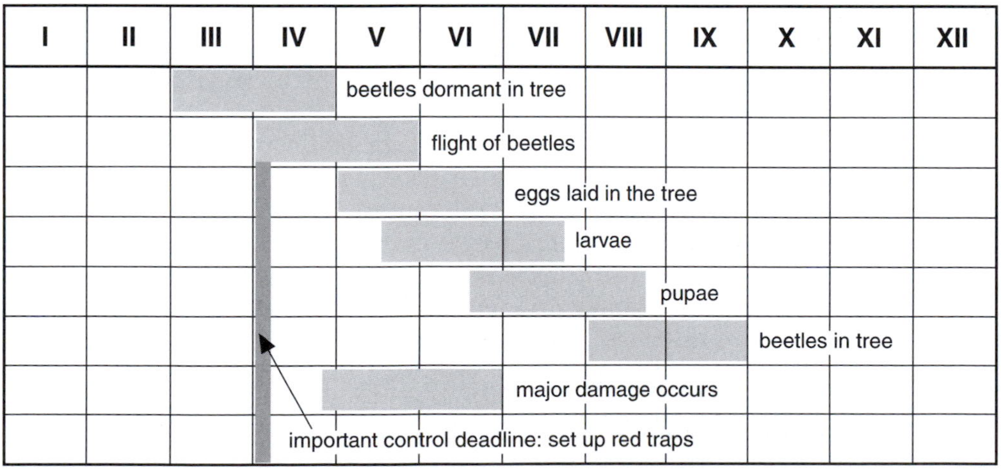

I	II	III	IV	V	VI	VII	VIII	IX	X	XI	XII
		____	beetles dormant in tree								
			____ flight of beetles								
				eggs laid in the tree							
				____ larvae							
					____ pupae						
							____	beetles in tree			
			____ major damage occurs								
			important control deadline: set up red traps								

Fig. 5.58. Life cycle of European shot hole borer (*Xyleborus dispar*).

From April until July the beetles can be seen on their host trees (apple, pear, cherry and plum), especially on hot, sunny days (temperatures above 18°C). After mating, the females bore into the bark of the host tree as far as the sapwood and make a cross-tunnel on each side. From this cross-tunnel, cylindrical brood tunnels are made, running upwards or downwards, and the eggs are laid at the entrance to these tunnels.

The larvae live from May to June in the brood tunnel in symbiosis with ambrosia fungi. Development into a beetle is completed in that same year. Both beetles and larvae of various ages can be found in the tunnels throughout the year.

NATURE OF DAMAGE.   From mid-May to June, holes with wood dust can be found in the older wood. The holes are continued in a widely ramified radial network of passages. Yellowish, brown-headed larvae or brownish-black beetles can be found inside the passages.

ENEMIES.   None.

INSPECTION AND DAMAGE THRESHOLD.   The trunk must be inspected for entry holes. The flight of the beetles can be monitored by means of red traps.

CONTROL

PREVENTIVE.   *Xyleborus dispar* mainly attacks young and frost-damaged trees. The beetle can smell frost-damaged trees from a considerable distance, and a very clear dividing line in the infestation is often discernible in the orchard (frost-damaged trees are attacked, while neighbouring healthy trees are unaffected).

The most important preventive measure, therefore, is the removal of frost-damaged and diseased trees.

Particular care should be taken with young orchards. Even in autumn, early frosts can often cause damage to the trees, especially if the ground is waterlogged. In the following spring these trees are attacked preferentially, particularly if the weather conditions are good.

DIRECT.   Spraying against *Xyleborus dispar* is not possible, but many of the beetles can be captured by setting up eight ethanol-baited red board traps per hectare. The smell of frost-damaged trees (due to fermentation processes in the cambium) is always more attractive, however, than the smell of the ethanol bait. There is no point in using this method, therefore, unless damaged trees are removed.

Apple sawfly *(Hoplocampa testudinea)*

APPEARANCE.   **Sawfly**: 6–7 mm long, brown or black with yellow-orange legs.
Egg: 0.8 mm long, white, oblong and slightly curved.

Larva: head initially black, later yellowish brown; body whitish.

Fig. 5.59. *Xyleborus dispar* boring a tunnel.

LIFE CYCLE

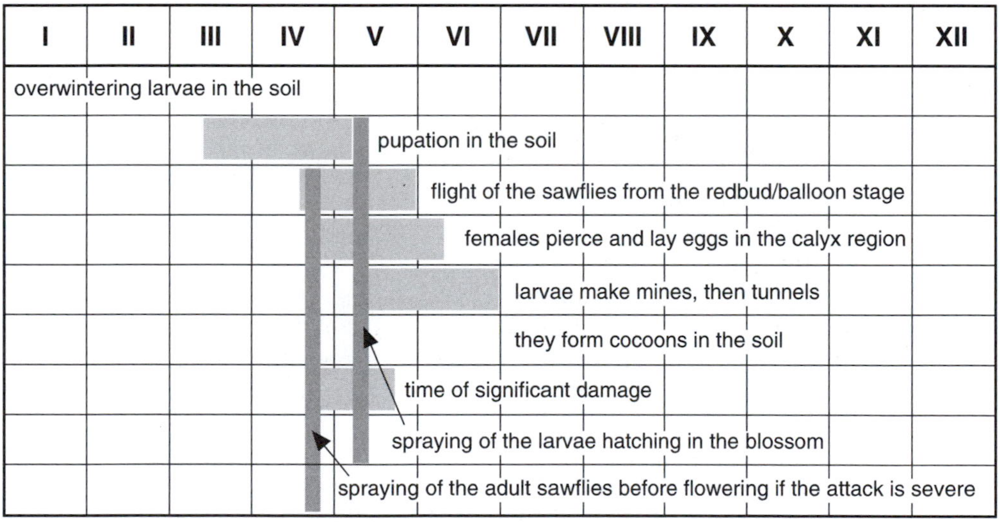

I	II	III	IV	V	VI	VII	VIII	IX	X	XI	XII
overwintering larvae in the soil											
			pupation in the soil								
				flight of the sawflies from the redbud/balloon stage							
				females pierce and lay eggs in the calyx region							
				larvae make mines, then tunnels							
				they form cocoons in the soil							
				time of significant damage							
				spraying of the larvae hatching in the blossom							
				spraying of the adult sawflies before flowering if the attack is severe							

Fig. 5.60. Life cycle of apple sawfly (*Hoplocampa testudinea*).

The apple sawfly overwinters in the larval stage in the soil. Some of the larvae overwinter for a further year, but most of them pupate in the soil. After about 3 weeks, the young sawflies hatch. The flight of the sawfly begins in the redbud stage at the earliest, but increases considerably in the balloon stage. The extent to which individual varieties are attacked by the sawfly is related to the order in which they flower. Early-flowering varieties are attacked more severely, late-flowering varieties less severely. After copulation the females lay their eggs in a slit in the base of the calyx. Only one egg is laid per blossom. The slit can be seen as a reddish brown patch in the area of the calyx. After about 2 weeks the larvae hatch and initially burrow just below the surface. Later the larva moves to the second fruit, where it makes a tunnel from which wet frass is emitted. The larva usually leaves this fruit through the widened entry hole and burrows into other fruits in the same blossom truss. Fruits in which a tunnel has been made will fall.

The apple sawfly larva can be distinguished from the larvae of the codling moth and fruitlet mining tortrix (which occur later) by the number of abdominal legs (eight pairs) and by its bug-like smell.

After 3–4 weeks, larval development is complete, and the larva makes a parchment-like, earth-coloured cocoon at a depth of 10–15 cm in the soil, where it overwinters.

NATURE OF DAMAGE

The damage caused can be seen as a small red dot where the slit has been made for oviposition in the calyx region of the blossoms.

The first damage seen in the fruit is a curved mine. Further fruits that are attacked exhibit a tunnel with wet, foul-smelling frass.

Early incisions made by sawflies in the calyx region can later be found as indentations in the calyx.

Fig. 5.61. Apple sawfly.

ENEMIES.   Since the egg and the larva are relatively well protected in the fruit, the main enemies are organisms living in the soil. Nematodes and fungi which parasitize the larvae overwintering in the soil are important. The degree of parasitization fluctuates from year to year.

INSPECTION AND DAMAGE THRESHOLD.   The flight of the apple sawfly can be observed very clearly with white traps of the Rebell type. Since these traps work by reflection of light, they must have maximum exposure to sunlight (so they need to be hung in a high, unobstructed position). The traps should already be hung up by the redbud stage. The damage threshold is 20–50 sawflies/trap.

It is absolutely essential to use the traps in early flowering varieties, as these are attacked first.

In order to monitor the hatching of the larvae, buds that have been attacked must be carefully dissected with a needle and the egg inspected. When the egg reaches the black-head stage, the larvae can normally be expected to hatch within the next 1–3 days. This time is critical if control measures are to be applied to the larvae as they hatch.

CONTROL

PREVENTIVE.   No preventive measures are known.

DIRECT.   If the damage threshold is exceeded, the hatching larvae should be sprayed with Quassia. To do this, however, it is essential to monitor the time of hatching very precisely (it usually coincides with petal fall in the individual varieties).

Since the sawfly attacks different varieties in the order in which they flower, different treatment has to be given to them. Early-flowering varieties are treated first, and then intermediate-flowering varieties. The degree of attack is usually very slight in late-flowering varieties.

A pyrethrum–rotenone spray should be used against the adult sawflies before flowering (when flight increases and the first slits are found) if there has been very severe sawfly infestation the previous year and large numbers of indentations have been found in the calyx region.

PEAR SUCKERS

Psylla pyri

Psylla pirisuga

Psylla piricola

APPEARANCE.   **Adult**: dark insect, 2–4 mm in size, with wings folded over the body.
Larva: 1–2 mm, bug-like, yellowish red or greenish grey in colour, with light transverse lines.
Egg: 0.6 mm long, oval, initially lemon-yellow, then orange just before hatching.

LIFE CYCLE

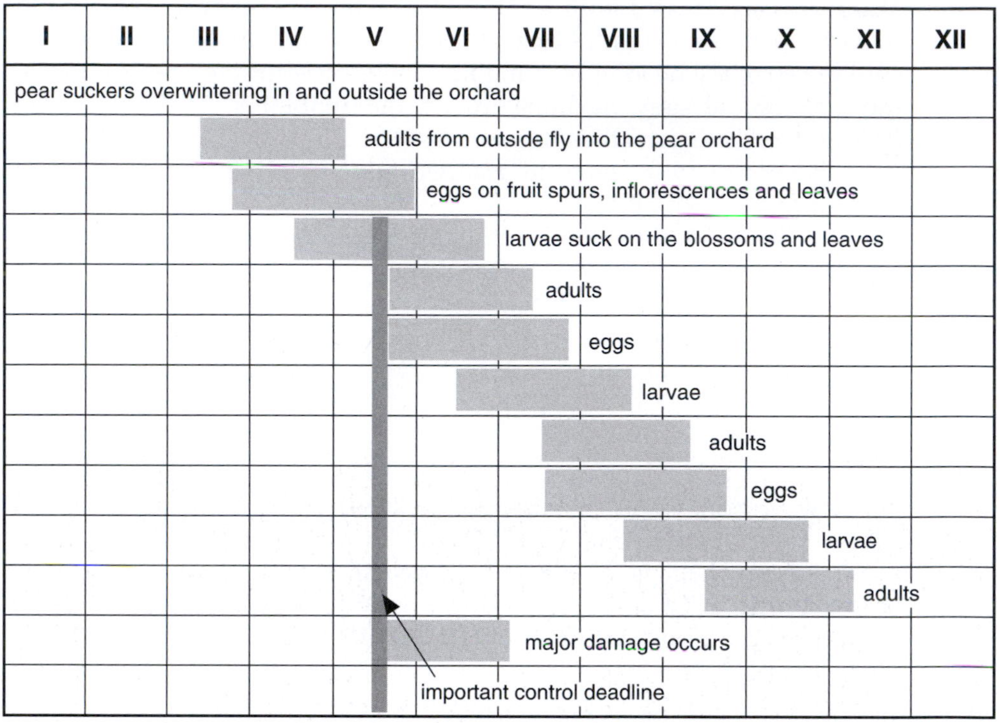

I	II	III	IV	V	VI	VII	VIII	IX	X	XI	XII
pear suckers overwintering in and outside the orchard											
			adults from outside fly into the pear orchard									
				eggs on fruit spurs, inflorescences and leaves									
				larvae suck on the blossoms and leaves									
					adults								
					eggs								
						larvae							
						adults							
						eggs							
							larvae						
							adults						
						major damage occurs						
				important control deadline								

Fig. 5.62. Life cycle of pear suckers (*Psylla* spp.).

The pear sucker species overwinter both in and outside the orchard. On the first warm days in March they emerge from their hiding-places and mate. At the end of March they start to lay their oval, yellow eggs, first on the fruit spurs and later on the inflorescences and leaves. The first larvae appear before flowering and begin immediately to suck on leaves and blossoms. From the third larval stage onwards they start to secrete honeydew. Two or three generations develop. Particularly between the end of May and middle of June the pear sucker population can increase dramatically or else completely collapse because of natural factors (predators and/or weather). Newly hatched pear suckers are very sensitive to cool, damp weather.

When the pears are so big that they point downwards rather than upwards, pear suckers like to get between two fruits and suck within this protected space. Sooty mould fungi, which blacken the fruit, shoots and leaves, grow on the honeydew that they secrete.

Psylla pirisuga causes hardly any damage. It only causes leaf deformations and moves away after flowering.

NATURE OF DAMAGE. Yellow-orange eggs are seen on the shoots in March and April. The bug-like larvae can be found on the leaves, shoots and axils, surrounded by sticky honeydew. Black patches are later found on fruit, leaves and shoots.

ENEMIES.   The most important enemy of the pear sucker is the anthocorid bug *Anthocoris nemoralis.*

This predatory bug overwinters as an adult in various hiding-places. Around flowering time it lays its eggs on trees with aphid or pear sucker colonies. In June the second-generation adults in turn look for suitable trees with pest colonies to lay their eggs on. This is usually the critical time in the organic control of the pear sucker. Natural regulation of pear suckers is very difficult if too many pear suckers are present or if they produce a lot of honeydew as a result of wet, sultry weather. If, however, enough of the whitish anthocorid bug eggs are found with a cap sunk in the leaf, regulation can occur after a few days.

In recent years there has been natural regulation of pear suckers by **hoverfly larvae** in June. The hoverfly appears to develop better in cold, wet periods than the anthocorid bug. In knockdown tests and visual inspections, 80% of the predators were hoverflies.

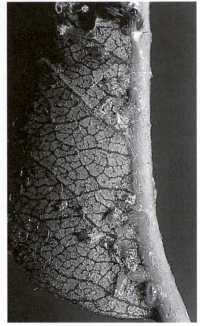

Fig. 5.63. Adult pear sucker.

Fig. 5.64. Pear sucker larvae: they secrete honeydew, which protects them.

INSPECTION AND DAMAGE THRESHOLD.   **First flowering**: visual inspection of the blossom trusses for leaf suckers and anthocorid bugs at any stage of development.

Second flowering: inspection of the shoots (especially the leaf axils) for leaf suckers and anthocorid bugs at any stage of development.

The damage threshold is very difficult to determine, as it depends on the anthocorid bug population.

The damage threshold has definitely been exceeded when pear sucker larvae are on the fruit.

CONTROL

PREVENTIVE

- The most important control measure is conservation of anthocorid bugs.
- The pear sucker is a pest which attacks the tips of shoots. Any measures which can check the growth of pear trees have a positive effect on infestation by pear suckers in the second and third generation.

DIRECT. The pear sucker is a pest which can very effectively be kept under control by organic control measures. In years with poor weather conditions in June and July, however, the development of anthocorid bugs may be too slow. There may then be an explosive increase in the pear sucker population.

The following measures should be applied if pear suckers have got between the fruits or if inspections give the impression that the anthocorid bugs are no longer able to create a stable situation:

- In orchards with overhead irrigation this should be switched on, so as to dissolve some of the honeydew.
- Soap or wetting agent formulations dissolve the honeydew and also have a limited toxic effect on the pear sucker larvae. Pyrethrum sprays also only have a limited effect on the larvae. Wait for rain or water the orchard before using a pyrethrum spray, so as to dissolve the honeydew.

Pear leaf blister mite *(Phytoptus pyri)*

APPEARANCE.   **Adult female**: 0.2–0.24 mm long, whitish to light brown, body narrow and elongated.

LIFE CYCLE

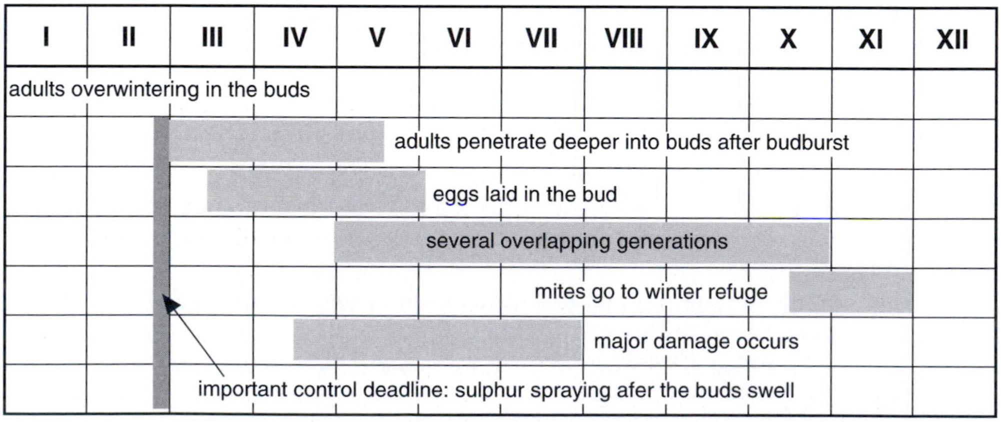

I	II	III	IV	V	VI	VII	VIII	IX	X	XI	XII
adults overwintering in the buds											
			adults penetrate deeper into buds after budburst								
			eggs laid in the bud								
				several overlapping generations							
					mites go to winter refuge						
				major damage occurs							
		important control deadline: sulphur spraying afer the buds swell									

Fig. 5.65. Life cycle of pear leaf blister mite (*Phytoptus pyri*).

Males and females overwinter beneath the outer bud scales. When the buds swell in the spring they become active, penetrate deeper into the bud and start to suck on the inner bud scales, where they lay their eggs. Later the mites feed on the leaves and the developing blossom buds. Initially the mites produce blisters on the underside of the leaf. After the tissue beneath the surface blisters has died, the mites can penetrate into the leaf. There the mites create pocket-like galls, where they live in the summer and where reproduction takes place. The young mites spread over the young shoots and cause new blisters there. In autumn, the fully grown juveniles move to their winter refuge in the buds.

NATURE OF DAMAGE.   The petals first turn red as a result of being sucked by the pear leaf blister mite. The yellowish-red blisters can be clearly seen when the first rosette leaves appear. The fruit is also damaged if infestation is severe, with extensive corky areas and sometimes malformations of the fruit.

ENEMIES.   No effective enemies of the pear leaf blister mite are known.

DAMAGE THRESHOLD.   No damage threshold is known. The level of infestation in the previous year is the only criterion that can be used. The pear leaf blister mite can show explosive development from one year to the next.

CONTROL.   If infestation by pear leaf blister mite was detected in the previous year, two sulphur sprays should be applied when the buds swell:

- first spray when the buds start to swell (7 kg/ha);
- second spray about 1 week later.

Pear midge *(Contarinia pyrivora)*

APPEARANCE.   **Midge**: midge 2.5–3.5 mm in size, with two longitudinal stripes.
Larva: 3–5 mm in length, yellowish white.

LIFE CYCLE.   The pear midge hibernates as a larva in a cocoon in the soil. After pupation in the spring the midges begin to fly from the mouse-ear stage onwards and lay their eggs on the stamens and pistil. The young larvae burrow into the ovary and feed on the internal part of the pear. Infested fruitlets initially grow faster and become bloated. After about 6 weeks the larvae leave the fruitlets, which are still hanging on the tree or have already fallen off, and move to their winter refuge in the soil.

NATURE OF DAMAGE.   Infested fruitlets grow faster after flowering and are sometimes misshapen and reddish coloured. Later on, growth ceases, the fruitlets turn black, sometimes crack open and drop off.

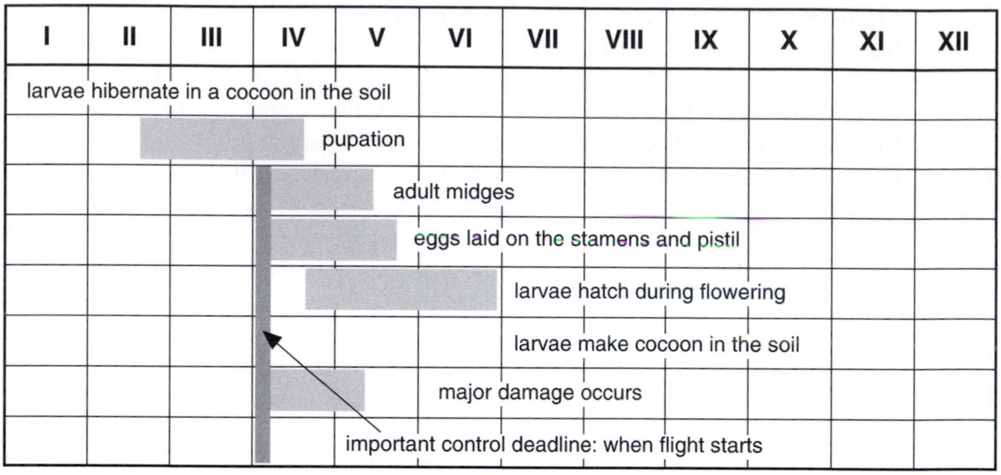

I	II	III	IV	V	VI	VII	VIII	IX	X	XI	XII
larvae hibernate in a cocoon in the soil											
				pupation							
				adult midges							
					eggs laid on the stamens and pistil						
						larvae hatch during flowering					
						larvae make cocoon in the soil					
					major damage occurs						
				important control deadline: when flight starts							

Fig. 5.66. Life cycle of pear midge (*Contarinia pyrivora*).

ENEMIES.   Chalcid wasps are the most important enemies.

INSPECTION AND DAMAGE THRESHOLD.   Flight and oviposition should be monitored on the blossom clusters. There is no known damage threshold.

CONTROL

PREVENTIVE

- Collect infested fruitlets and remove them from the orchard.
- Keep hens in the orchard (hens eat the larvae).

DIRECT.   The value of control measures depends on the level of infestation in the previous year. Treatment should be carried out if there was severe infestation in the previous year and blossom production is slight.
     Pyrethrum–rotenone sprays have some effect.

Water vole (Arvicola terrestris)

The water vole is now an endangered species in Britain but in many countries in mainland Europe it is considered an agricultural pest.

APPEARANCE.   The water vole is somewhat smaller than a rat, and has a shorter tail and a blunt nose. It can grow to a length of up to 25 cm (including the tail).

LIFE CYCLE.   The water vole lives underground, is exclusively herbivorous and feeds mainly on fresh roots (unlike the field mouse, which feeds mainly on the less succulent parts, especially the roots of the M9 apple rootstock and the elder). It makes a complex system of tunnels in the ground, reaching a length of up to 80 m and usually going down to a depth of 50 cm.

The cross-section of the tunnels is a vertically elongated oval shape, unlike the tunnels made by moles, which have a horizontally elongated oval shape. The hills formed by the soil that is excavated are usually flat and located to the side of the hole, whereas molehills are higher and located directly over the hole. Plant parts are also often found in the excavated material. The vole very rarely leaves the tunnel.

The succulent roots, tubers and bulbs preferred by the vole are usually stored in large quantities in storage chambers. In a dry, well-lined nest the female gives birth to 2–6 young, 3–4 times a year from March to October. Since the vole is sexually mature after only 2 months, there can easily be a population explosion. Dry weather, in particular, encourages vole reproduction, whereas in cold, wet weather the nests get very wet and many young voles perish.

NATURE OF DAMAGE. In spring the trees produce few or no shoots. Trees attacked can be easily shaken, and if the roots have already been eaten away the tree can simply be pulled out of the ground.

During the growing season the trees initially begin to wither, then later they die.

ENEMIES. Although the water vole has many natural enemies, it can still cause a great deal of damage.

Among birds of prey the most effective enemy is the buzzard. The population of this predator, however, is very small, and since the water vole is hardly ever encountered above ground, the total beneficial effect of the buzzard is very slight. Owls are another natural enemy of the water vole. Among domestic animals, cats and dogs are important predators of the water vole, with some dogs being just as effective as cats.

Another enemy, the weasel, is found mainly in quiet places, near water.

In the UK, a major enemy is the mink, which is thought to have caused a serious decline in the water vole population.

INSPECTION AND DAMAGE THRESHOLD. In mainland Europe the damage threshold for the water vole is zero. As soon as a water vole is detected in an orchard, it should be captured.

CONTROL. The water vole is a key pest in organic apple and elder cultivation in some areas of mainland Europe (e.g. the Styrian region of Austria). Organic production in these areas cannot be successful unless the water vole is kept under control.

PREVENTIVE

- Encourage beneficial predators by creating perches for birds of prey and piles of stones for weasels.
- Cut grass regularly.
- Good mechanical tilling of the row of trees, especially in autumn, when it is overgrown with weeds.

- Do not let wild plants grow freely (wild-flower strips and extensive meadows) in orchards that are at risk of attack from water voles.
- The roots are often surrounded by wire netting when the tree is planted. This wire netting only provides protection in the first year and possibly in the second year. After that the roots have already grown out through the netting.

DIRECT. The only really effective method of controlling water voles is to catch them in traps.

First of all, the tunnel must be found and a 'burrowing test' made. In this test the tunnel is opened over a length of about 50 cm, in such a way that it follows a straight course. If the tunnel is closed again within the next hour, this means that the burrow is still inhabited. The tunnel which has been closed is opened up again, and the trap, disguised by rubbing earth over it, is put into the tunnel. The tunnel is then closed in such a way that the trap is not visible.

A small cartridge-firing device, about 20 cm in length, can also be put into the tunnel instead of a vole trap. Compared with the trap this device has the advantage that it also kills voles which push a small amount of soil along in front of them, as the cartridge penetrates through the soil and kills the vole, whereas the trap would be blocked by the soil in front of the vole.

On the following day, at the latest, the tunnel can be carefully re-opened and the result checked.

Major diseases and pests of stone fruit

Virus diseases

Virus diseases cannot be controlled. The highest priority must therefore be given to ensuring that the planting stock is clean. Even so, re-infection often cannot be ruled out in the case of many virus diseases, e.g. plum pox in plums and apricots.

Table 5.14. Virus diseases of plums, peaches and apricots (after Lankes, 1996).

Causal agent	Disease/symptoms	Vector	Economic damage
Plum pox virus	**plum pox** leaf chlorosis scars on fruit cracks in bark	aphids	severe losses in yield if fruit cannot be marketed
Apple chlorotic leaf spot virus	**pseudopox** fruit symptoms similar to plum pox cracks in bark		grafting failures, shortened life of the trees

Continued

Table 5.14. *Continued.*

Causal agent	Disease/symptoms	Vector	Economic damage
Apple mosaic / other causes?	**yellow bud mosaic** mosaic-like light patches on leaves		grafting failures, reduced growth of young shoots
Prune dwarf virus	**blind wood** dwarf plums shortened internodes	pollen and seeds	grafting failures, dwarf growth, losses in yield
Prunus necrotic ringspot virus	**Prunus necrotic ringspot** leaf symptoms: necrotic ringspots, line or oak leaf pattern	pollen and seeds	dieback symptoms in some cultivar/rootstock combinations

Table 5.15. Virus diseases of cherries.

Causal agent	Disease/symptoms	Vector	Economic damage
Prune dwarf virus	**chlorotic ringspot disease** leaf symptoms	pollen and seeds	frequently latent (sweet cherries) growth and yield depression; increased PNRV symptoms
Prunus necrotic ringspot virus	**Stecklenberg disease** leaf, blossom and fruit symptoms enations	pollen and seeds	reduced shoot growth reduced leaf formation delayed start to flowering delayed ripening yield losses (40–90%)
Raspberry ringspot virus	**Pfeffing disease** narrow leaves enations bare patches	nematodes	in sweet cherries: fruits stay small, drop early dieback symptoms
Little cherry disease	small fruits fruits stay small and light-coloured leaf symptoms rare	mealybug	losses in yield and quality

Fungal diseases

Agents with fungicidal effect that are currently available to organic grow-ers are copper and wettable sulphur, as well as various plant tonics based on substances of plant origin (e.g. lecithin) or mineral origin (e.g. stone dust). Great importance is attached to prevention by using a 30% wider spacing than in conventional orchards, together with a moderate N supply and loose tree structure.

The economic damage threshold indicates the level of infection which is still just tolerable. This is exceeded when the fungal pathogen in question causes greater financial loss than it would cost to prevent the damage. From this is derived the control threshold. This is the level of infection at which control measures must be applied in order to prevent adverse economic effects.

Table 5.16. Major fungal pathogens of cherries.

Pathogen/main symptom	Evidence of infection When to check	Type of check	Control threshold	Remarks
Monilinia brown rot Fungus grows after the blossoms open through the flower stalk into the wood and leads to the well-known brown rot	appearance of first petals	–	not determined	risk of infection in cool wet weather. Cut out diseased shoots. Bear varietal differences in mind. Schattenmorelle especially susceptible
Cherry leaf spot disease	mid-May to mid-June duration of leaf wetness	check temperature, relative humidity and leaf wetness > 12 h	not determined infect. conditions: temp. > 17°C, relative humidity > 60%	risk especially in areas with high rainfall. Schattenmorelle especially susceptible
Gnomonia leaf scorch	dried-up leaf trusses in winter	count	not determined	if infection is severe: treatments at budburst and April to mid-May
Wood and bark diseases	gummosis, sunken bark, etc.	visual	not determined	cut out visible sites of infection. Control at budburst or at leaf drop

Fig. 5.67. Plums attacked by plum pox.

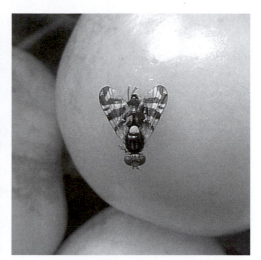

Fig. 5.68. Tunnel made by the plum fruit moth larva in the fruit.

Fig. 5.69. Adult fly laying eggs on the fruit.

Table 5.17. Major fungal pathogens of plums.

Pathogen/main symptom	Evidence of infection When to check	Type of check	Control threshold	Remarks
Monilinia brown rot	see cherries	see cherries	see cherries	see cherries
Pocket plums	malformed fruit in spring	count	not determined	Auerbacher, Bluefre, Ortenauer, President and Hauszwetschke are especially at risk. Control at budburst and in short first flowering
Plum rust	from mid-June to mid-July, yellow spots initially on upper side of leaf	count	not determined	control from end of June. Increased incidence usually only local. Especially susceptible are Auerbacher, Ersinger, Cacaks Schöne and Ruth Gerstetter
Wood and bark diseases	see cherries	see cherries	see cherries	see cherries

Animal pests

Table 5.18. Major animal pests of cherries.

Pest/main symptom	Evidence of attack When to check	Type of check	Control threshold	Remarks
Winter moth Caterpillars usually hatch from the winter eggs before flowering and devour leaf and blossom buds. Later caterpillar stages also damage the fruit	dormancy of growth before flowering	check branch samples visual check	2–3 eggs/1 m of fruit spurs 3–4 caterpillars per 100 blossom trusses	apply sticky bands in October to catch the females. Renew the glue in February/March. *Bacillus thuringiensis* sprays
Black cherry aphid Hatches from winter eggs at flowering time, forms colonies at tips of shoots. Also leaf curling in sweet cherries	second flowering	visual check	occurrence of the first colonies 2–5 colonies/100 shoot tips	control early, so as to prevent honeydew production. Aphid changes host plant. Summer host: bedstraw and speedwell
Cherry fruit fly Maggots burrow into the flesh of the fruit and feed near the stone. Pupation in the soil takes place after 3–4 weeks	yellow or yellowish red colouring of the fruits (from mid-June onwards)	monitor flight with yellow traps on south or south-west side of the trees	1–2 flies/yellow trap per day	varieties are especially at risk from the 4th week of maturity onwards. Eggs are laid only on the fruits that are changing colour

Table 5.19. Major animal pests of plums.

Pest/main symptom	Evidence of attack When to check	Type of check	Control threshold	Remarks
Winter moth Caterpillars usually hatch from the winter eggs before flowering and devour leaf and blossom buds. Later caterpillar stages also damage the fruit	flowering to second flowering	visual check	10–15 caterpillars per 100 blossom trusses	apply sticky bands from October onwards. Use green sticky band paper, as this catches fewer beneficials. Renew the glue in February/March

Continued

Table 5.19. *Continued.*

Pest/main symptom	Evidence of attack When to check	Type of check	Control threshold	Remarks
Plum fruit moth Caterpillar burrows into the flesh of the fruit ('red plum maggot'). Infested fruits ripen prematurely	May–June (1st generation) July–August (2nd generation)	visual check	1–3 fruits with eggs laid on them (always on the underside of the fruit) per 100 fruits	determine the peak of flight with pheromone trap. Advisable to check for eggs in order to determine the time for control measures. First generation rarely warrants control
Plum sawfly Enters directly below the sepals; fruits attacked drop off	about 1 week before start of flowering until end of flowering blossom	white trap to predict flight visual check	not determined 4–8 entry points per 100 fruit calyxes	level of the control threshold depends on blossom formation. Thinning desirable if crop load is heavy
Leaf-curling plum aphid Shoot growth is reduced, severe leaf curling, fruit is undersized	first flowering to second flowering (April/May)	visual check	1 colony/100 shoots 2–3% of buds infested with aphids	significant plum pox vector. Migrates to asters, chrysanthemums and clover from May onwards. Winged aphids return to the plum trees in autumn to lay eggs
Damson–hop aphid Large amounts of honeydew produced, resulting in growth of sooty moulds – fruit contaminated; aphid on underside of leaf; no leaf curling	flowering to second flowering	visual check	1 colony/100 shoots	occurs especially in warm years, migrates to hops in summer
Plum scale Large amounts of honeydew produced, resulting in growth of black moulds. Contaminated fruit is undersized and ripens prematurely	budburst until balloon stage	visual check	more than 100 live scale larvae per 2 m of fruiting wood. Take previous year's infestation into account	the scale insects have only one generation per annum. Control is possible with plant oils, e.g. rapeseed oil

Continued

Table 5.19. *Continued.*

Pest/main symptom	Evidence of attack When to check	Type of check	Control threshold	Remarks
Spider mites Yellowish patches at first on the leaves, starting from the base	winter dormancy, first flowering	check branch specimens	800 eggs per 2 m of fruiting wood	In organic orchards control is only possible at the egg stage with an oil spray
of the leaf, spreading over the leaf in a netlike pattern, greyish brown	flowering, second flowering	visual check	50% of larvae hatched (first mobile stages)	
discoloration, leaves curl in drought, shoot length and fruit size reduced if infestation is severe	summer until about 3 weeks before start of harvest	visual check	40–50% of leaves attacked (= 2–4 mites/leaf)	

Source: Rhineland Palatinate, *Guidelines for integrated fruit growing*, 1996, slightly amended.

It is virtually impossible to prevent fruit from becoming worm-infested. Although *Bacillus thuringiensis* sprays are effective in theory, the plum fruit moth remains on the wing for a very long time, and the larvae which are supposed to ingest the insecticide burrow very quickly into the fruit. Careful checks to determine flight peaks and the hatching of eggs are therefore necessary. As in the case of cherries, early varieties are less at risk.

Major diseases and pests of strawberries

Diseases

Grey mould *(Botrytis cinerea)*

NATURE OF DAMAGE.   Initially there are brown spots at sites of infection on unripe fruit; ripening fruits are covered with mouse-grey mould.

BIOLOGY

- Most important cause of damage in organic production; infection can lead to very big losses.
- Mould overwinters in plant debris (leaf stalks, mummified fruit).
- High relative humidity and temperatures of 15–20°C during flowering are conducive to the primary infection. Fruit becomes more susceptible with increasing ripeness. The disease can spread rapidly in wet weather.

CONTROL

PREVENTIVE

- Ensure good ventilation (wide spacing, single-row instead of double-row systems).
- Sparing use of nitrogen fertilizers.
- Do not grow for more than one season.
- In spring, remove dead leaves and fruit that has been infected during the harvest.
- Apply straw mulch at the right time.
- If the strawberries are watered, this should be done in the morning rather than evening.
- Aerate frequently if the strawberries are grown in tunnels.
- Do not use susceptible varieties.
- Infection is significantly reduced if the plants are covered with a rain shelter (low tunnel) during the harvest.

DIRECT. No effective fungicides are known at present.

Powdery mildew *(*Sphaerotheca macularis*)*

NATURE OF DAMAGE. **Leaves**: white mildew coating on the underside of the leaves, which subsequently turn reddish violet and curl upwards.
Flowers and fruit: development of flowers and fruit is retarded; a white mildew coating initially covers the seed grains, then the entire fruit.

BIOLOGY

- The fungus overwinters mainly on strawberry plants.
- Temperatures of 20–25°C are conducive to the fungus.

CONTROL

PREVENTIVE

- Ensure good ventilation (wide spacing, single-row instead of double-row systems).
- Do not use susceptible varieties.
- Sparing use of nitrogen fertilizers.
- Do not grow for more than one season.

DIRECT. Spray with wettable sulphur; if infection is severe, give preventive treatment before flowering; for strawberries grown over two seasons, give 0.3% treatment to the new growth, if infected, after pruning (concentration 0.2%) and repeat after a week.

Fig. 5.70. Grey mould rot in strawberries: brown spots initially, later covered with grey mould.

Fig. 5.71. Crown rot: when cut open, the rhizome shows reddish brown areas affected by rot.

Leaf spot and scorch

(*Mycosphaerella fragariae* and *Diplocarpon earliana*)

NATURE OF DAMAGE. **Leaf spot**: small, round spots with a brown outline and white centre (leaves).
Scorch: small, irregularly shaped brownish red spots without a white centre (leaves).
Other spots on leaves can be caused by fungi such as *Gnomonia comari*, *Colletotrichum acutatum* or *Alternaria alternata*.

BIOLOGY

- Rarely a problem in strawberries grown in a single season.
- Fungi overwinter on strawberry plants.
- High relative humidity is conducive to the diseases.

CONTROL

PREVENTIVE

- Do not grow for more than one season.
- Do not use susceptible varieties.
- Ensure good ventilation (wide spacing, single-row instead of double-row systems).

DIRECT. Copper sprays depending on incidence, 3 and 6 weeks after planting, at budburst in spring, for strawberries grown over two seasons, treat the new growth after pruning (concentration with 50% pure copper: 0.05–0.1%).

Angular leaf spot *(Xanthomonas fragariae)*

NATURE OF DAMAGE. Light-green, watery, angular spots initially on the underside of the leaf. Subsequently spots also become visible on the top of the leaf and turn black. Bacterial mucus is secreted on the underside of the leaf.

Sepals, blossoms, runners, leaf stalks and fruit stalks are also attacked.

Reduced growth in latently infected young plants.

BIOLOGY

- Becoming more important.
- Causes little damage in some years but considerable damage in others.
- Bacteria survive up to 2 years on dead plant material.
- Daytime temperatures around 20°C, cool nights and high relative humidity are conducive to infection.
- Warm water treatment of young plants encourages the spread of the disease.

CONTROL

PREVENTIVE

- Use healthy plant material.
- Do not share or exchange machinery with plantations affected; workers should change their clothing and wash their hands after working in infected plots.
- Sparing use of nitrogen fertilizers.
- Aerate frequently if the strawberries are grown in tunnels.

DIRECT. No effective formulations are known at present.

Red core *(Phytophthora fragariae)*

NATURE OF DAMAGE

- Symptoms easiest to see in spring, and sometimes also in autumn.
- Poor budburst and stunted growth in spring, sometimes death of entire plants, little or no fruit formed.
- Older leaves pale or reddish brown in colour, with shortened stalks.
- When the root is cut through lengthwise, from healthy to diseased tissue, a red central core with a lighter root bark is discernible.
- Main roots without lateral roots ('rat tails').

BIOLOGY

- Can lead to heavy losses.

- The development and spread of the fungus is increased if the soil is compacted, waterlogged or variable in moisture content.
- Diseased stocks can partially recover during the summer.
- Spread by diseased plant material, soil tillage equipment, footwear and soil water (especially significant on slopes).
- The pathogen persists in the soil for at least 15 years.

CONTROL

PREVENTIVE

- Do not grow on soils that are compacted, waterlogged or variable in moisture content.
- Use healthy plant material.
- Wait at least 15 years before replanting infected plots with strawberries.
- Do not exchange or share machinery with affected plantations.
- Do not grow for more than one season.
- Do not use susceptible varieties.

DIRECT. No effective fungicides are known at present, and they are hardly needed if the preventive measures are borne in mind.

Rhizome rot and leather rot (Phytophthora cactorum)

NATURE OF DAMAGE. **Rhizome rot**: symptoms of damage usually appear a few weeks after planting or soon after flowering. The plant wilts from inside, the leaves then turn brown, and the plant dies. When cut open lengthwise, the rhizome shows reddish brown areas affected by rot, which are clearly differentiated from the healthy tissue.
Leather rot: a brown discoloration is found on unripe fruits; the fruits have a leathery consistency. The symptoms of damage (watery white, brown infected areas, flesh bitter) are rarer in ripe fruits.

BIOLOGY. **Rhizome rot**: waterlogged soils and temperatures above 25°C are conducive to infection, with early-harvested strawberries being especially at risk.
Leather rot: infection occurs through infected soil getting on to fruit via irrigators.

CONTROL

PREVENTIVE

- Do not grow on soils that are compacted, waterlogged or variable in moisture content.
- Use healthy plant material.

- Apply straw mulch at the right time.
- Do not use susceptible varieties.
- Do not grow after phacelia or papilionaceous plants (*Fabaceae*).
- Remove diseased plants at an early stage.

DIRECT. No effective fungicides are known at present.

Pests

Strawberry blossom weevil (Anthonomus rubi)

NATURE OF DAMAGE. Severed flower buds, which dry up and then fall off. **Beetle:** blackish brown, 2.0–3.5 mm long; segmented antennae; wings dotted in a longitudinal direction.

BIOLOGY

- Beetle overwinters beneath leaves (sometimes blown in from woodland) and straw or in the soil.
- One female destroys up to 20 flower buds, by laying an egg in each of them.
- Young beetles feed on the foliage after hatching, without causing much damage.
- Slight infestation in strong-flowering varieties has a desirable thinning effect.
- Possibility of confusion with the rarer strawberry stem weevil (*Coenorhinus germanicus*); the latter, however, does not have segmented antennae and severs entire flower stems.

CONTROL

PREVENTIVE

- Do not grow on plots near to woodland.
- If there is infestation, grow only strong-flowering varieties.
- Covering with fleece (early-harvested strawberries) offers some protection against leaves blown in from woodland.

DIRECT. No effective insecticides are known at present.

Strawberry mite (Tarsonemus pallidus)

NATURE OF DAMAGE

- Mites barely visible to the naked eye (about 0.2 mm in length).
- From July onwards the very young leaves remain small, curl up and turn brown.

- The attack spreads rapidly from a focal point.
- Can be confused with the symptoms of damage by leaf nematodes, but the latter is already visible in the spring.
- Rarely a problem in single-cropping varieties.

CONTROL

PREVENTIVE

- Use plant material that is free of infestation (treated with warm water if necessary).
- Do not grow for more than one season.

DIRECT.   Destroy infested plants.

Major diseases and pests of raspberries

Diseases

Root rot (Phytophthora fragariae *var.* rubi)

NATURE OF DAMAGE.   **New canes**: Tips of shoots wilt, leaves become paler and then dry up, shoots die by early summer.
Fruiting canes: production of stunted fruit shoots, leaves become paler and then dry up, shoots die around harvest time or earlier.
Roots: bark dark coloured, few rootlets.

BIOLOGY

- Leads to losses which endanger the crop.
- The development and spread of the disease is increased if the soil is compacted, waterlogged or variable in moisture content.
- Greatest risk of infection in spring and autumn at soil temperatures of 12–16°C.
- Spread by diseased plant material, soil tillage equipment, footwear and soil water (especially significant on slopes).
- Strawberry plants are not attacked.

CONTROL

PREVENTIVE

- Do not grow on soils that are compacted, waterlogged or variable in moisture content.
- Use healthy plant material (use foliage plants instead of canes or pot plants).

- Regularly add well-rotted compost, combined with ridge culture.
- Water sparingly in spring and autumn.
- Do not exchange or share machinery with affected plantations.
- Do not use susceptible varieties.
- Wait at least 15 years before replanting infected plots with raspberries.
- Use rain shelter from beginning of March until end of November.

Raspberry cane diseases (Didymella applanata, Leptosphaeria coniothyrium, Botrytis cinerea, Elsinoe venta)

NATURE OF DAMAGE. *Didymella applanata*

- Violet-brown patches, which rapidly increase in size, around buds of the new canes.
- Silvery-grey discolorations with small black perithecia in the winter, sometimes discoloration of the entire canes.
- Buds open poorly or do not open at all.

Leptosphaeria coniothyrium

- Large violet-brown patches at the base of the new canes.
- Infected canes die in the following year.

Botrytis cinerea

- Pale-brown patches, which rapidly increase in size, around the buds of the new canes.
- Silvery-grey discolorations with large black mould patches in winter.
- Buds open poorly or do not open at all.

Elsinoe venta

- Initially crimson, then greyish white sunken patches on canes, leaf stalks and blades.

BIOLOGY. Unlike root rot, cane diseases never lead to the death of the whole plant.

CONTROL

PREVENTIVE

- Avoid damage to canes, e.g. by suppressing the raspberry cane gall midge or by removing too strong canes with torn bark.
- Remove fruited canes immediately after harvesting.
- Use nitrogen fertilizers sparingly.
- Ensure the canes are well spaced, i.e. remove surplus new canes and weeds.

Fig. 5.72. Raspberry cane disease.

Grey mould *(*Botrytis cinerea*)*

NATURE OF DAMAGE. Fruits covered by powdery mouse-grey fungal growth; the berries become soft and rotten, then subsequently shrivel up and become hard.

BIOLOGY

- Causes considerable damage if there is wet weather during harvesting.
- Apparently healthy fruits can become rotten after a short period in storage, especially if they are wet when picked.
- The fungus overwinters on canes (raspberry cane diseases).
- The principal infection occurs during flowering.

CONTROL

PREVENTIVE

- Use a rain shelter.
- Use nitrogen fertilizers sparingly.
- Remove infected canes in winter.

Pests

Raspberry beetle *(*Byturus tomentosus*)*

APPEARANCE AND NATURE OF DAMAGE. **Beetle**: 3.5–4.5 mm in length; brown to greyish brown, covered by flat-lying hairs. It hollows out flower buds and feeds on flowers and young leaves.
Larva: 6–8 mm in length; yellowish brown; feeds on the basal drupelets.

BIOLOGY. The main damage is caused by larvae (infested fruits).

- Entire consignments can become unsaleable because of infestation.
- The presence of the beetles can be detected by means of white sticky traps (which should be put in place in mid-April).

CONTROL

PREVENTIVE

- Choose autumn-bearing varieties instead of summer-bearing varieties (autumn varieties are hardly attacked at all).
- Do not plant in plots near woodland.

Major diseases and pests of blackberries

Diseases

Blackberry tendril disease *(Rhabdospora ramealis)*

NATURE OF DAMAGE. **New canes**: dark-green, pinhead-sized spots at the base of the shoots; the spots then turn reddish and later brownish with a red outline. When the spots get bigger, some of them merge and spread on the upper parts of the tendrils.
Fruiting canes: leaves and flowers wilt and dry up from the end of the tendril towards the base of the shoot. The tendrils above the areas attacked die. Pycnidia, out of which whitish fungal tissue grows in damp weather, usually develop in rows in spring. The spots then fade from the centre outwards.

The disease is often mistaken for frost damage, but in the latter there is no damage to new canes and no pycnidia can be seen in spring.

CONTROL

PREVENTIVE

- Do not use susceptible varieties (Theodor Reimers is susceptible).
- Remove and burn badly infected tendrils.

Pests

Blackberry mite *(Acalitus essigi)*

NATURE OF DAMAGE. Ripening berries with red drupelets that stay hard and sour.

BIOLOGY

- Mites become active from March onwards.
- Level of infestation increases in the course of the harvest.
- Mites are not visible to the naked eye (0.16–0.2 mm in length).

CONTROL

PREVENTIVE.   If possible, remove fruited canes in the same year (**Caution:** this increases the risk of frost damage).

DIRECT

- Treatment with 2% wettable sulphur if the lateral shoots of the fruiting canes are 10 cm in length.
- Treatment with 1% wettable sulphur if the lateral shoots of the fruiting canes are 20 cm in length.

Major diseases and pests of currants and gooseberries

Diseases

American gooseberry mildew (Sphaerotheca mors-uvae)

NATURE OF DAMAGE
In gooseberries:

- Shoot tips and young leaves covered by white fungal mycelium.
- Shoot tips die.
- Production of broom-like new shoots begins.
- Fruits covered initially by a white mycelium, then by a rough brown 'scab-like' coating.

In currants: as gooseberries, but there is hardly any infection of fruit.

BIOLOGY

- Fungus overwinters on shoot tips and infected buds.
- Can be mistaken for European gooseberry mildew (*Microsphaera grossulariae*), but the latter causes very little damage and usually appears after the harvest, mainly on the underside of the leaf.

CONTROL

PREVENTIVE

- Do not use susceptible varieties.
- Use nitrogen fertilizers sparingly.
- Remove and destroy infected shoot tips.

- **Before budburst:** treatment with 0.5% wettable sulphur or 0.4% fennel oil.
- **After budburst:** treatment with 0.4% fennel oil every 2–3 weeks, depending on the severity of infection.

Caution: do not spray into open flowers (risk of failure to set fruit); many varieties are sensitive to sulphur after budburst.

White pine blister rust (Cronartium ribicola)

NATURE OF DAMAGE

In currants:

- More and more orange spore colonies on the underside of the leaf from June onwards.
- From July onwards, progression to 'pillars', 1–1.5 mm in length.
- Premature leaf drop.

In white pine:

- Fusiform swelling of parts of the branches or trunk.
- Yellowish white fungal tissue then breaks through the bark.

BIOLOGY

- Host-changing fungus: goes through various developmental stages on five-needled pine species, the most important of which in central Europe appears to be the white pine (*Pinus strobus*).
- The fungus can presumably also survive without an intermediate host (*Pinus* sp.).
- Other rust fungi apart from *C. ribicola* also occur on the *Pinus* species in question.
- The disease is most severe on blackcurrant but also occurs on redcurrant varieties derived from *Ribes petraeum* and on *R. alpinum* and *R. grossularia*.

CONTROL

PREVENTIVE

- Do not use susceptible varieties.
- Remove infected *Pinus* species in the immediate vicinity.

DIRECT. From budburst onwards, treatment every 2–3 weeks with 0.4% fennel oil, depending on the severity of infection.
Caution: Do not spray into open flowers, because of the risk of failure to set fruit.

Pests

Various aphid species

NATURE OF DAMAGE

- Colonies develop mainly on shoot tips.
- Shoot and leaf malformations.
- Purple blisters on red- and white currants, yellowish green blisters on blackcurrant (currant blister aphid: *Cryptomyzus ribis*).

BIOLOGY. Aphids often cause damage not only by sucking but also by carrying dangerous virus diseases.

CONTROL

PREVENTIVE

- Use nitrogen fertilizers sparingly.
- Create favourable conditions for beneficials (e.g. by planting wild-flower strips near the crop).

DIRECT. Use pyrethrin, rotenone, fatty acid or composite sprays before shoot or leaf malformations occur.

Blackcurrant gall mite *(Cecidophyopsis ribis)*

NATURE OF DAMAGE. **In blackcurrant**, infested buds are rounded and swollen (galled buds), fail to open normally, and dry up.
In redcurrant and gooseberries the damage is as in blackcurrant, except that the buds do not swell up but are merely somewhat looser.

GENERAL ASPECTS. Mites transmit reversion and increase the risk of failure to set fruit. There are differences in susceptibility between varieties.

CONTROL

PREVENTIVE

- Do not use susceptible varieties.
- Remove and destroy infested buds or shoots before flowering starts.

Currant clearwing moth *(Synanthedon tipuliformis)*

APPEARANCE. **Moth**: Wing span about 20 mm, body bluish black, abdomen with four (male) or three (female) bright yellow bands.
Caterpillar: 23–30 mm long; yellowish white with a brown head.

Fig. 5.73. Currant clearwing moth.

NATURE OF DAMAGE. Caterpillars burrow into the shoots in June–July and make a black tunnel in the pith. The result is that leaves wilt, shoots fail to open properly, and some of them die.

CONTROL

PREVENTIVE

- Consistently rejuvenate the bushes; any summer pruning should not be carried out until after the moths take wing, from August onwards.
- Remove and destroy severely infested shoots.
- Use traps to reduce infestation (e.g. wine bottles or bark beetle traps, available commercially). The traps should be baited with 90% cider, 5% blackcurrant cordial and 5% vinegar and should be spaced at intervals of about 20 m. The bait should be regularly renewed if contaminated.

Major diseases and pests of bilberries

In some areas of the world a number of diseases and pests can cause severe damage to bilberries. Anthracnose – presumably caused by *Colletotrichum acutatum* – appears to play a central role. As yet, little is known about this disease, but there is some evidence that microclimate, cultural techniques and choice of variety are critical factors determining whether infection occurs.

Problems can be prevented by using nitrogen fertilizers sparingly, making sure that plantations are well ventilated, setting up rain shelters if necessary in areas with high rainfall, and taking care not to make plantations too big.

Pesticides

The most important pesticides used by organic fruit growers are briefly discussed in this section. National guidelines and conditions have been disregarded in listing the pesticides, as they change very rapidly.

The organic pesticides are natural substances that have little toxicity.

Fungicides

Copper

Table 5.20. Copper.

Areas of application	Scab, fruit tree canker
Mode of action	Preventive, kills the germinating spores
Advantages	Is highly resistant to rain, has very good redistribution on the tree, highly effective against scab
Disadvantages	Causes severe russeting; copper leads to severe russeting of fruit if applied from flowering until the June fruit drop. The degree of fruit russeting depends on the amount of copper and of water. Fine spraying causes less phytotoxic damage than normal spraying. Copper can also lead to phytotoxic damage on the leaves, however. Copper is not broken down in the soil Copper is toxic to earthworms
Amount used	The amount used must not exceed 1.5 kg of pure copper/ha per annum, i.e. about 2.5–3 kg of the final product In very critical periods: 0.5 kg/ha In less critical periods: 0.3 kg/ha

Sulphur

Table 5.21. Sulphur.

Areas of application	Mildew, scab, rust mites, pear pox mites
Mode of action	Preventive against fungi, kills the germinating spores
Advantages	Sulphur also has an acaricidal effect. Control of mildew in spring also controls rust mites Highly effective against mildew

Continued

Table 5.21. *Continued*

Disadvantages	Temperature-dependent (at least 15–18°C) At higher dosages (over 3 kg/ha) and higher temperatures sulphur is harmful to predatory mites Very poor effect against scab Is not tolerated by all apple and pear varieties Cannot be mixed with paraffin oil
Amount used	So as not to harm predatory mites, not more than 3 kg/ha 7 kg/ha against the pear pox mite

Calcium polysulphide

Table 5.22. Calcium polysulphide.

Areas of application	Against overwintering stages of scale insects on the tree in dormancy, and against scab
Mode of action	Corrosive action against overwintering stages, preventive action against scab – inhibits the germinating spores
Advantages	Very effective against scab Good adhesion (similar to copper) Does not cause russeting
Disadvantages	Is not currently registered for use against scab Phytotoxic damage at too high dosages Cannot be mixed with paraffin oil
Amount used	15–20% as a winter spray (in absolute winter dormancy) 10–20 kg/ha against scab

Acaricides and insecticides

Paraffin oil

Table 5.23. Paraffin oil.

Areas of application	San José scale, red spider mite
Mode of action	The pests are suffocated by the film of oil
Advantages	Very effective against San José scale and red spider mite, does not harm beneficials
Disadvantages	Not biodegradable There should not be any frost for 2–3 days after spraying (otherwise there is phytotoxic damage or in susceptible varieties, such as Gala and Braeburn, the blossom trusses and leaves fall)
Amount used	10–30 l/ha, depending on stage of development
Remarks	Optimum time for use is the mouse-ear/redbud stage It is absolutely essential to spray each row twice (e.g. 2 × 15 l if the amount used is 30 l/ha) Fine spraying gives better coverage Rapeseed oil is much less effective than paraffin oil but is easily degradable

Granulosis virus

Table 5.24. Granulosis virus.

Areas of application	Codling moths (Carpovirusine, Madex, Granupom) Tortrix moths, *Adoxophyes orana* (Capex)
Mode of action	Ingestion insecticide
Advantages	Highly infectious and toxic Highly specific effect Does not harm predatory mites
Disadvantages	Is rapidly broken down by UV radiation, so only effective for a few days under natural conditions
Remarks	The timing of sprays must be very flexible, depending on weather Slight damage may be caused because the caterpillars are not killed immediately

Bacillus thuringiensis *ssp.* kurstaki *(Bactospeine, Dipel, Thuricide)*

Table 5.25. *Bacillus thuringiensis* ssp. *kurstaki* (Bactospeine, Dipel, Thuricide).

Areas of application	Winter moth, fruitlet mining tortrix, noctuid moths
Mode of action	Ingestion insecticide
Remarks	Only works well at temperatures above 15°C, as not enough is ingested at low temperatures, because the larvae do not feed actively Addition of 1% sugar improves the effect Not harmful to beneficials

Azadirachtin (neem)

Table 5.26. Azadirachtin (neem).

Areas of application	Woolly apple aphid, winter moth
Mode of action	Repellent and ingestion insecticide Inhibits development (sloughing) and reproduction
How obtained	Neem seed oil extract (alkaloid)
Remarks	Very slow action, so spraying to control woolly aphid needs to be carried out before flowering Only permitted in nurseries at present Dangerous to bees Also effective against tortrix moths at very high concentrations

Quassia

Table 5.27. Quassia.

Areas of application	Apple and plum sawfly, also effective against aphids, tortricid larvae and cherry fruit fly
Mode of action	Repellent, contact and ingestion insecticide against sawfly
Advantages	Not harmful to beneficials
Remarks	Use against sawfly at end of flowering

Pyrethrum–rotenone sprays (Parexan)

Table 5.28. Pyrethrum–rotenone sprays (Parexan).

Areas of application	Aphids, blossom weevils and other weevils, tortricids, winter moth, bugs, noctuid moths
Mode of action	Contact, ingestion and inhalation insecticide
How obtained	Pyrethrum is produced from chrysanthemums and contains piperonyl butoxide as an additive which increases UV stability
Remarks	Very broad spectrum of action Not completely harmless to beneficials Pyrethrum is poisonous to fish Sprays consisting of pyrethrum alone are also commercially available (e.g. Spruzit)

Ryania

Table 5.29. Ryania.

Areas of application	Codling moths, tortrix moths, aphids, leaf miners
Mode of action	Contact insecticide
How obtained	Ryania is obtained from the wood of *Ryania speciosa*
Remarks	Broad spectrum of action More persistent than pyrethrum–rotenone sprays Poisonous to fish Not registered in the EU at present

Soft soap insecticides, fatty acids

Table 5.30. Soft soap insecticides, fatty acids.

Areas of application	Aphids, pear suckers
Mode of action	Direct insecticidal action is only moderate, dissolves honeydew
Remarks	Russeting can occur under dry weather conditions Spraying at apple blossom time has a thinning effect

6 Processing

Under EU Regulation 2092/91, not more than 5% of the ingredients used in organic produce (as a percentage by weight at the time of processing) may be of non-organic origin. All other ingredients must comply with the requirements of organic farming. This applies in particular to the fruits used, which must originate from an orchard that is recognized as organic.

General requirements for fruit for processing

As a general principle, a high-quality final product can only be made from a high-quality primary product. Four minimum requirements can therefore be laid down for fruit to be processed.

Fruit to be processed must be **ripe**, **healthy**, **clean** and **suitable**.

Ripe fruit

The riper the fruit, the more flavour it has. Unripe fruit usually has little flavour, and sometimes even poor flavour. Because of the low sugar content of unripe fruit, products made from it do not have enough fullness of flavour. In this type of fruit, the ratio of sugar to acid is not right, and products made from it do not have a harmonious flavour.

Something similar applies to overripe fruit. It has usually gone beyond its flavour peak and is already decomposing. For this reason fruit which is in storage should not be processed. The right degree of ripeness is generally reached when the fruit is suitable for eating fresh.

© CAB International 2003. *Organic Fruit Growing* (K. Lind, G. Lafer, K. Schloffer, G. Innerhofer and H. Meister)

Healthy fruit

This means that fruit which is rotten or mouldy should never be sent for processing. Apart from an insipid, putrid flavour, this type of fruit contains a large number of unwanted microorganisms (yeasts, moulds and bacteria) and thus a lot of enzymes. This increased microbial count can subsequently lead to difficulties in clarification, pasteurization or fermentation.

Clean fruit

Fruit for processing is often fruit that has fallen from the tree or bush. It is usually heavily contaminated with leaves, grass, stones or soil (1 ml of soil contains more than a thousand million microorganisms).

The fruit should therefore be cleaned before processing if possible.

Suitable fruit

Depending on the desired final product, individual fruits may be more or less suitable for processing. Not all species (or varieties) of fruit are suitable for all products.

Production of fruit juices

Preparing the fruit

Sorting and cleaning

After the fruit has been received at the factory it has to be cleaned. In practice, however, only pome fruit is washed – this is rarely done with stone fruit and berry fruit. Cleaning is completed by respraying the fruit on the elevator or by counter-current rinsing with process water in a spiral conveyor. In addition to visible contaminants, microorganisms are also removed in the washing process, so the microbial count is significantly reduced. If both organically and conventionally produced fruit is processed in the factory, care should be taken to ensure that residues do not get onto the organic fruit via the water used for washing. Fresh water must therefore always be used for cleaning organic fruit.

It is absolutely essential to remove all rotten, damaged and/or unripe fruit. At present this is done manually, on endless conveyors or roller conveyors.

Stemming and stripping

Stemming is very important in the processing of stone fruit (cherries). The green stalks contain tannins which can get into the product during processing and cause a disagreeable flavour. Stemming can be omitted in the processing of pome fruit.

Adapted grape processing equipment is generally used for stripping. A low proportion of combs makes it easier to press out the mash. If the proportion is too high, unwanted tannins get into the juice.

Crushing the fruit

The method and degree of fruit crushing have a decisive effect on juice removal. The greater the degree of crushing, the greater is the number of cells damaged. If the fruit is crushed very small it is much more difficult to separate the juice from the solids, and a high suspended solids content in the juice increases clarification costs. Care should be taken to ensure that the degree of crushing is as uniform as possible.

During the crushing process, ascorbic acid can be added to protect against oxidation. This prevents non-enzymatic browning until the juice is pasteurized.

In many cases mash pumps are used to pump the crushed fruit to the juice extractors, mash heaters or other equipment, resulting in further comminution.

In some cases the processes are combined with the use of enzymes (mash fermentation). A pitting machine can be inserted in the stone fruit processing line ahead of the crusher.

Mechanical processes

Various kinds of fruit mills are now commonly used for crushing fruit. They are usually made from stainless steel and plastic, because these materials are resistant to the aggressive acids in the fruit and are easy to clean.

Table 6.1. Different types of fruit mills.

Type of mill	Suitable for	Characteristics
Grinding mills	Pome fruit	Different grinding knives are used, depending on the fruit to be processed
Centrifugal mills	Pome fruit	The degree of crushing is determined by the hole diameter
Wing roller mills	Stone and berry fruit	The roller spacing can be varied and the mill thus adapted to the fruit to be processed
Hammer mills	Pome fruit	Of little significance
Scraper mills	Pome (and stone) fruit	Mainly small-scale use

Thermal processes

Thermal processes (heating, thermobreak or freezing) are only used in exceptional cases and are generally insignificant.

Pre-treatment of the mash

Heating the mash

Enzymatic pectinolysis is often performed before the crushed fruit is pressed, in order to obtain satisfactory juice yields and facilitate pressing. This processing step takes place at a higher temperature level, as enzymes only work slowly at low temperatures. Steam-heated heat exchangers (mainly tubular heat exchangers but also spiral heat exchangers or heat exchangers with rotary scrapers) are generally used for heating the crushed fruit.

Equipment for heating the crushed fruit must meet the following requirements:

- continuous operation
- the flow should be as turbulent as possible in order to ensure rapid heating of the crushed fruit
- it should take up little space
- it should be easy to put together and to clean.

Treatment with pectolytic enzymes

Most commercial products are a mixture essentially of four individual enzymes: pectin esterases, polygalacturonases, pectin lyases and pectate lyases. Depending on the intended use, the enzyme preparations have different contents of the individual pectolytic enzymes. The enzyme preparations are commercially available in liquid or solid form; they originate mainly from mould cultures.

As a general rule, when using enzymes it is important to make sure that they have not been obtained from cultures of microorganisms that have been altered by genetic engineering.

When crushed apples are treated with pectolytic enzymes the temperature should not exceed 30°C, as otherwise there is likely to be a change in aroma quality (ester-type components of the apple juice aroma, in particular, may be destroyed). It is recommended that the juice be left for about an hour at this temperature. The minimum temperature of the juice for efficient enzyme use is 10°C, but at this temperature the time for which the juice needs to be left has to be trebled in order to achieve approximately the same effect as at 30°C. The enzyme should be added while in the mill. **The yield can be increased by up to 10% by using special enzymes.**

Berry fruit is usually treated with enzymes at 50–55°C. This is virtually the optimum temperature for the enzymes. After about 2–4 h there are significant increases in pressability and also in colour yield. If the mash is highly acidic, care should be taken to choose enzymes that are still sufficiently active at low pH.

In the case of berry fruit, the enzyme should already have been added before the mashing process, as otherwise it is difficult to achieve a uniform distribution in the mash. Mashes which are stirred for a long time at elevated temperatures become mushy and the fine suspended solids content increases markedly. This may subsequently lead to considerable difficulties during clarification.

Mash transport

The mash is conveyed from the fruit mill to the press, and to and from the heat exchanger, by means of valveless eccentric spiral pumps, large impeller pumps or slow-running piston pumps.

The pumps need to be correctly dimensioned in terms of their capacity, as fast-running pumps cause additional crushing of the mash particles, especially if the material has been enzymatically treated, and this increases the suspended solids content of the juice. The resultant fine suspended solids are difficult to remove, thus increasing the clarification costs.

Juice extraction

Most of the technical procedures employed are based on the same pressing process that has always been used. The overriding concern in juice extraction is to avoid oxidation – which makes rapid extraction essential. The juice extraction equipment must therefore satisfy the following criteria.

From the point of view of maintaining quality, the extraction process must be rapid and exclude air, so as to prevent changes in quality. To this end the plant should be technically efficient, if possible continuously operable, and highly reliable, with few breakdowns. It should also require few staff to operate it. The plant should provide a maximum yield, run economically and be easy to clean.

The following factors are determinative in the pressing process:

- pressure
- degree of crushing
- prior extraction of juice
- layer height.

The pressure, degree of crushing, etc., will of course vary from case to case and depend on the quality of the fruit and the experience and skill of the operator.

Pressing aids are devices or substances by means of which it is possible to improve the structure of the material to be pressed, the internal surface and thus the extraction of juice during the pressing process. The additives needed (principally cellulose fibres or perlite) are in the range 0.5–1.0% by weight, irrespective of the pressing aid employed. They make it easier to extract the juice from the remaining material. The presses most commonly used in fruit processing are the rack and frame press, the hydraulic horizontal basket press and the filter belt press.

Rack and frame press

The rack and frame press is a further development of the basket press. Rack and frame presses are commonly used on farms that process apples and pears. They are particularly suitable for mashes that are difficult to press. They give high yields and good juice quality with little suspended material and tannin.

In terms of their design they are fairly small in size, but the output is high. The biggest disadvantage of the rack and frame press is the high labour requirements. Other disadvantages are the severe oxidation of the juice produced and the irregular flow of juice.

Hydraulic horizontal basket presses – Bucher-Guyer system

The design of horizontal basket presses has major advantages over the earlier vertical basket presses.

In newer designs the press is filled via a central pipe. The bottom of the press can already be revolving while it is being filled, and in this way uniform filling of the press is achieved. Filling takes place under pump pressure and can therefore be automated. This results in better prior extraction of the juice and also a higher filling capacity.

It is possible to increase the yield by incorporating drainage tubes. These are covered with a kind of pressing cloth and make it possible for the juice to run off internally. In this way it is possible to achieve a significant reduction in the distance travelled by the juice, resulting in a shorter pressing time. In the closed cylinder there is complete exclusion of air during extraction of the juice, so oxidation is well nigh impossible.

The removal of the residue can also be automated. This means that one man can operate two or more presses.

Filter belt presses

Belt presses can be continuously fed with mash. Their output is high, ranging from 2 to 6 t/h. Modern filter belt presses achieve high yields. Because of their rapid operation (juice extraction times of about 4 min) there is little oxidation.

Between the belts large shearing forces act on the mash. The suspended solids content is therefore somewhat higher than in other pressing processes. In order to reduce these shearing forces, belt presses with only one belt have been built in recent years. This allows the suspended solids content of the juice produced to be significantly decreased.

Other types of presses

Table 6.2 lists the other types of presses also used in fruit processing.

Table 6.2. Different types of fruit presses.

Type of press	Characteristics
Vertical basket press	Juice travels a long distance; low yields; high labour requirements; hardly used at all nowadays
Mechanical horizontal basket press	Mainly used for berry fruit (wine grapes); high space and labour requirements; little used nowadays
Hydro press	High oxidation; high water consumption in spite of small size; only for small quantities
Pneumatic presses	Mainly used for berry fruit (wine grapes)
Screw presses	Continuous operation possible; high output; few breakdowns; high tannin and suspended solids content; severe oxidation

Other methods

For many years mechanical pressing was the only method employed. Alternative and/or auxiliary methods were developed, however, because the need to ensure continuous operating cycles, the high labour costs, changes in raw materials, unsatisfactory yields and other factors made it essential to find new approaches in juice extraction technology.

JUICE EXTRACTION USING WATER.   In this process the juice is extracted from the fruit with water. In large-scale trials it was possible to achieve yields of over 95% of the total extractable substances in the fruit. This process is of major significance in factories where the juice obtained is further processed to a concentrate.

There are significant sensory and analytical differences between juices obtained by pressing and those obtained by extraction with water. The mineral content of juices obtained by pressing is about 10% lower on average. Their polyphenols content is also considerably lower, sometimes only half that of juices obtained by extraction with water.

ENZYMATIC LIQUEFACTION OF THE RAW MATERIAL.   Mixtures of pectinolytic and cellulolytic enzymes are used. Together they bring about a stepwise degradation, with initial loosening and separation of the cell aggregates, then breakdown of the cell walls and finally conversion of cellulose to sugar. After liquefaction of the mash, the remaining solids are usually separated off in a decanter.

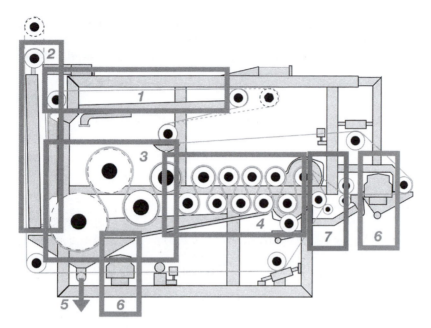

Fig. 6.1. Twin-belt filter press (Bellmer): 1. Prior extraction of juice, 2. Wedge zone, 3. Low-pressure pressing zone, 4. High-pressure pressing zone, 5. Juice runoff, 6. Filter cleaning, 7. Filter cake removal.

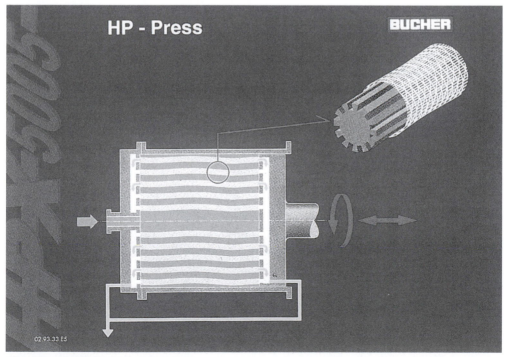

Fig. 6.2. Hydraulic horizontal press.

Production of clear fruit juices

The most common process for production of apple juice is shown in Fig. 6.3.

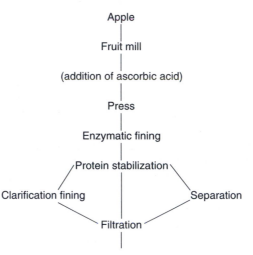

Fig. 6.3. Common processes in the production of apple juice.

The clarification of juices, i.e. the mechanical removal of suspended solid particles from the juice, is generally carried out by **sedimentation**, **filtration** or **centrifugation**. These processes are very often combined in order to achieve a better result.

Sedimentation

Suspended solids settle in the tank because of differences in density. This process can be speeded up by clarification fining or by reducing the viscosity of the juice (using enzymes).

Enzymatic fining

ENZYMATIC HYDROLYSIS OF PECTIN.   In many factories this fining process is carried out at a juice temperature of 50–55°C. After only a few hours the juice is virtually clear (often completely clear) and can be processed further.

When this fining process is applied to juices that are not heated, it is usually combined with clarification fining. The enzyme then has a longer time to act, which is needed because of the low temperature. **There is no point in using pectolytic enzymes at a juice temperature below 10°C.**

The enzymes used for juice clarification are almost exclusively pectolytic. The reasons for the clarification effect are as follows. Pectin stabilizes the suspended solids in the juice. Through addition of the enzyme, the pectin is converted to a dissolved form and broken down. This is associated with a significant decrease in viscosity. In addition, pectins that have been made soluble act as protective colloids for many of the suspended particles. As soon as the protective effect of pectin is lost, these suspended particles coagulate and are precipitated. A significant decrease in viscosity is accompanied by a significant increase in filtration performance.

The suspended particles mainly consist of protein and have a negatively charged surface. Precipitation of the suspended solids can therefore be speeded up by adding colloids that are positively charged at this pH, such as gelatin.

In organoleptic studies no significant changes in odour and taste could be found when the commercially available enzyme preparations were used even at concentrations many times the normal level.

ENZYMATIC HYDROLYSIS OF STARCH. Starch is a substance used for storing energy reserve in plants, where it is deposited in the form of starch granules. Starch may be noticed in fruit juice as a milky haze. This haze disappears when the juice is heated, but re-appears after it is cooled to less than 10°C. Like pectin, starch has a protective colloid effect on suspended particles and thus makes juice clarification more difficult.

Hazes caused by starch mainly occur when unripe or prematurely ripe fruit is processed, but can be avoided by preventive use of enzymes that hydrolyse starch.

Clarification fining

Fining of juices generally removes not only unwanted substances but also, to a certain extent, desirable substances. For this reason excessive fining is to be avoided. One aim is to cause as little change as possible in nutritional and organoleptic properties.

The effect of fining is optimal when the fining agent is added continuously. In this way excessive concentrations in certain places are avoided and the fining agent is uniformly distributed in the juice.

The more acidic the beverage, the better is the clarification effect with a given level of fining agent. The result is also temperature-dependent. For successful fining, the temperature of the juice should not normally be less than 12°C.

Considerable amounts of suspended material accumulate in the fining of fruit juices – 3–10% as a rule, but even higher percentages are possible. The quantity of this material depends on various factors. In large factories, treatment to remove it is absolutely essential. Various methods are available for this purpose.

The most common methods of treatment are the chamber filter press or rotary vacuum filters. The latter method is the one generally used today.

GELATIN FINING.   There are several reasons for using gelatin. Firstly, in clarification fining it rapidly coagulates with negatively charged particles (suspended solids and/or silica sol), and secondly it forms compounds with phenols and precipitates them as well. In addition to flavour changes due to removal of tannins, the latter are no longer able to form compounds with protein or heavy metals which might be perceptible as a haze. Brown-coloured polyphenols also react with gelatin, so the fining process results in a lighter colour.

The optimum amount of gelatin is determined in a preliminary trial. If the amount added is too high, the bottled juice may be subject to hazes which appear under cold storage conditions but disappear again at higher temperature. Moreover, an excessive level of gelatin results in significant flavour loss.

The success of gelatin fining essentially depends on the temperature of the juice. At low temperature (less than 7°C) the gelatin does not flocculate completely (so hazes may develop later). The process should therefore be carried out at temperatures around 15°C.

SILICA SOL–GELATIN FINING.   Silica sol is an aqueous, colloidal solution of silicic acid. It has a milky, opalescent appearance. Only negatively charged silica sol is used for fining purposes. It is marketed in the form of a 15% or 30% solution. Silica sol is never used alone in fining, but essentially only as a reactant with gelatin (or other positively charged flocculants such as casein). In addition to being used as a reactant, it also prevents excessive concentrations of gelatin. Addition of silica sol in fining is particularly useful if gelatin fining is impossible because the polyphenol content is too high or too low.

In the fining of juices with a low tannin content, the silica sol is added before the gelatin, but in the case of juices with a high tannin content the gelatin is added first (in powder form) and then silica gel (30%), with about 5–7 times as much silica sol as gelatin being used. The amount of silica sol used is determined by the amount of gelatin used.

TANNIN–GELATIN FINING.   Tannins are tanning agents which are readily soluble in water, and which have different chemical compositions, depending on their origin. Tannins can be used as an aid in gelatin fining.

This type of fining is of little significance nowadays.

Protein stabilization

Most fruit juices contain thermolabile protein which can cause hazes in the final juice. These hazes are undesirable, so the thermolabile protein is removed from the juice during processing. Two methods are commonly used for this purpose.

BENTONITE FINING.   Bentonite fining is primarily used to stabilize fruit juices against protein hazes. Bentonites are swelling clays. They have a lamellar structure that, depending on the type of bentonite, can absorb different amounts of water of crystallization and different amounts of exchangeable cations (calcium, magnesium or sodium) between the layers. The swelling capacity of bentonites varies, depending on origin and composition: sodium bentonites have a high swelling capacity, while calcium bentonites have a low swelling capacity. Composite products (sodium–calcium bentonites) are often offered for sale. Because of differences in charge properties, bentonite reacts with proteins and tannins. In addition, bentonite treatment reduces the levels of heavy metals and any spray residues present.

The amount of bentonite needed for fining (about 0.5–3.0 g/l) is determined by preliminary trials. If too much is used, both the colour and the flavour of the product are impaired.

The required quantity of bentonite is pre-swelled with about 5–10 times as much water. When cold water is used, the bentonite must be left longer in the water (about 8–10 h) than when warm water is used (3–4 h). After swelling is completed, the supernatant water can be poured off and the bentonite suspended with some of the juice. Pre-swelling has also been found to be beneficial in the case of preparations that are already granulated. This suspension is then mixed into the juice, stirring constantly. The juice must be kept in motion for at least a 15 min. The fining has no effect without suitable mixing and a minimum juice temperature of 12°C.

HIGH-TEMPERATURE SHORT-TIME (HTST) METHOD.   The HTST method is used for two purposes in fruit juice technology: to sterilize juices for storage and to precipitate thermolabile protein.

Within a short time the juice is heated to temperatures between 82 and 90°C, depending on the product. This temperature is maintained for a short time, and the juice is then quickly cooled in order to avoid serious losses of quality. Thermolabile protein is denatured in the hot phase and coagulates. In the cooled juice this suspended material is precipitated after a few hours.

Separation

In the fruit juice industry, separators are used primarily for removal of suspended solids. In the separator cloudy juice is subjected to a high centrifugal force, thus separating the solids from the liquid.

Because of the solids content and the particle size of the solids, semi-closed or closed (hermetic), self-discharging plate separators are best suited for the clarification of fruit juice.

Filtration

Filtration means the removal of suspended solids from the juice by means of a porous layer (consisting of a filter aid or filter aids) which is permeable to the juice but impermeable to the solids. When cloudy juice flows under pressure through a porous layer, the suspended solids are retained on the surface (surface filtration) or in the interior of the layer (depth filtration).

In surface filtration the solids retained are those that do not pass through the smallest cross-section of the capillary flow channels of the filtering layer. Many particles are trapped by adsorption in the labyrinthine three-dimensional sieve of the filter aid. This means that substances can be retained that are smaller than the mesh size of the filter aid. In depth filtration, on the other hand, the solids are trapped in the interior of the layer; this is due to the mechanical retention capacity (inertia and size of the particles, sedimentation, diffusion) and to the composition of the juice.

Filter aids must meet the following requirements:

- chemical stability and physiological safety
- optimum particle size and shape
- favourable price.

SHEET FILTRATION.   Sheet filters consist of a tubular steel chassis or a free-standing filter chassis in which square metal or plastic filter plates are vertically arranged. The filter plates are fluted plates made of ribbed material or tubular frames with perforated plates.

The arrangement of the inflow and outflow, respectively, gives rise to turbid and clear chambers alternately. Layers consisting of cellulose, diatomaceous earth and perlite are usually employed for the filtration of fruit juices. The degree of clarification by the individual sheets varies, depending on the amount of diatomaceous earth used, and ranges from coarse filtration to sterilizing filtration (sterilizing sheets). Sterilizing sheets are only used in special cases in fruit juice production, as fruit juices are usually given heat treatment to prolong their shelf life. Sterilizing filtration is therefore not needed to ensure the keeping quality of the product.

The operation of filter sheets is based on three factors:

- **Surface filtration**: coarse particles are trapped on the surface because of their size.
- **Depth filtration**: removal of fine particles and microorganisms in the pores of the sheet.
- **Adsorption effect**: microorganisms and suspended particles are retained on the fibres of the filter sheets because of differences in electric charges.

Before a sheet filter is used for the first time, the filter sheets should be rinsed for about 15 min with water, to make sure that the product does not have any off-flavour due to the sheets (paper or diatomaceous earth flavour). For this rinsing operation the flow rate should be high and the filter sheet pack should not be tightened.

Under normal circumstances the flow rate of the liquid to be filtered in the sheet filter should not be too high, as otherwise the mechanical and physicochemical retention capacity is insufficient to guarantee effective clarification. The flow rate to be chosen depends on the configuration of the sheets. If the speed at which the liquid arrives is too great it 'bursts through' the sheet, nullifying the filtration effect. The performance of a sheet filter depends on the size and number of the sheets and on the degree of turbidity or the nature of the sheets.

With increasing filtration time, the particles separated from the juice clog up the capillaries in the sheets. In surface filtration the very nature of the operation inevitably means that the surfaces of the sheets get clogged up in the course of filtration.

Both these factors together raise the flow resistance, and there is a consequent decline in filtration performance. The performance of the sheet filter is limited by the capacity of the sheets.

PRECOAT FILTRATION – DIATOMACEOUS EARTH FILTRATION. The most economical method of filtering fruit juices with a high suspended solids content is with precoat filters. In these, the filter sheet is created by precoating a liquid-permeable filtering element with the filter aid. In addition to diatomaceous earth, perlite and cellulose are also used as filter aids in fruit juice production.

The amount of diatomaceous earth used depends on the turbidity of the juice. Highly turbid juices require significantly more diatomaceous earth than less turbid products. The performance of the filter depends on the area and number of the filtering elements and the turbidity of the juice.

The equipment most commonly used has vertical or horizontal filtering elements.

Table 6.3. Comparison of vertical and horizontal filtering elements.

Horizontal	Filtering elements	Vertical
possible	uniform precoating of the filter sheet	not possible
not possible	bilateral precoating of the filter sheet	possible
possible	interruptions or replacement of the filter medium	difficult
high	space requirements (for a given filter area)	low
possible	automatic discharge	not possible
possible	complete evacuation	not possible

DIATOMACEOUS EARTH FRAME FILTER. This type of filter is usually found in smaller factories. The diatomaceous earth frames are arranged as filter plates in a carrier chassis. They fall within the group of vertical precoat filters.

The frames are normally about 40 mm wide. Re-usable carrier sheets, which are used as carriers for the filter cake, can be used here to reduce the weight. The advantage of this type of equipment is that the frame is also suitable for filter plates for sheet filtration.

MEMBRANE FILTRATION.   Membrane filter cartridges are seldom used in fruit juice technology, as hot-filling is the common practice and prior membrane filtration is therefore unnecessary. These filters can only be used after fine filtration, as otherwise the membrane is immediately clogged up.

Production of naturally cloudy juices

The cloudy juice contains colloids which increase the viscosity of the juice (and prevent sedimentation) while at the same time binding suspended particles by virtue of electric charges. These colloids which are responsible for the haze stability of a juice are mainly pectins and starch. It is therefore a key priority in processing to retain these colloidal substances. The unripe fruit contains a lot of starch but little soluble pectin, whereas the fully ripe fruit contains little starch and a lot of soluble pectin. This means that it is possible to influence the haze stability of the juice by selection of the raw material.

In order to retain the light colour, ascorbic acid can be added to the juice immediately after pressing. If some of the fruit used is unhealthy, the ascorbic acid should already be added in the mash, to prevent browning during processing.

Four methods are widely used at present for production of naturally cloudy juices.

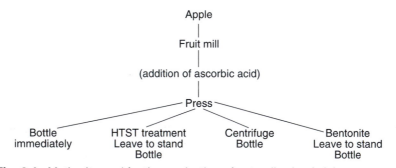

Fig. 6.4. Methods used for the production of naturally cloudy juices.

Further processing without being left to stand

The freshly pressed juice is collected in a bulk tank, from which bottling and pasteurization take place. Apart from the press and the pasteurizer, there is relatively little need for technical equipment. Particles of widely varying size are suspended in the juice. Coarse particles are precipitated soon after pasteurization and form an unwanted sediment in the bottle.

Since no suspended solids are removed in this method, it is especially important to choose a pressing process which leaves little suspended material.

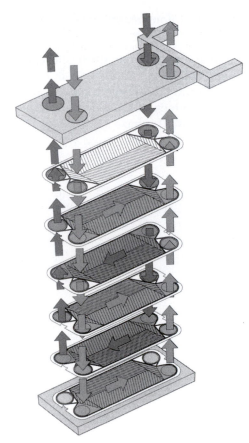

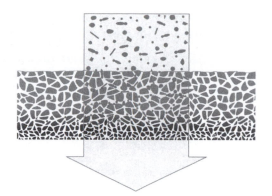

Fig. 6.6. Principle of depth filtration (BEGEROW).

Fig. 6.5. Flow diagram for plate heat exchanger (TETRA PAK).

Further processing after bentonite treatment and being left to stand

The freshly pressed juice is treated with bentonite (see section on *Bentonite fining*) and left to stand for 6–8 h. During this period, coarse particles and the bentonite are precipitated. Bottling and pasteurization then take place. The enzymes can break down pectin during the period while the juice is left to stand. This can lead to partial clarification of the juice in the bottle.

Further processing after centrifugation

In this method the juice goes from the press into a bulk tank. From this it passes through a centrifuge, in which the coarse particles are eliminated, depending on the speed of rotation. Suspended solids of uniform fineness are left in the juice. There is only a short lapse of time between the press and bottling, so pectolytic enzymes have no time to take effect.

Further processing after high-temperature short-time treatment and then being left to stand

After leaving the press the juice is given high-temperature short-time treatment. Protein is denatured by the high temperature and is coagulated. After being left to stand for a few hours the protein and the coarse suspended solids are precipitated. The enzymes of the juice are inactivated by the heat treatment, this being essential for haze stability.

Extending the shelf life of fruit juices

In practice pasteurization is virtually the only procedure used to give fruit juices a longer shelf life.

Pasteurization

AIMS OF PASTEURIZATION

- To inactivate all the enzymes contained in the juice.
- To kill all the microorganisms contained in the juice.
- To kill all the microorganisms present in the bottle.

Pasteurization is of great importance for juice quality. The recommended bottling temperature for clear apple juices is 78°C, while naturally cloudy juices should be bottled at a temperature of at least 80°C. Too high temperatures or too long a holding phase lead to partial conversion of fruit sugars (fructose and/or glucose) into hydroxymethylfurfural (HMF). This process is known as 'caramelization'. An excessively high HMF content is referred to as 'cooked flavour' and is regarded as a fault.

Bottles which immediately after hot-filling are stacked in large storage containers (e.g. pallet boxes) retain the heat for a very long time. Although such juices have a fairly reliable shelf life, there is a significant loss of quality because of the prolonged action of the heat.

Faults in pasteurization are the main reason for deficient keeping quality of fruit juices. Simple thermometers can show divergences of up to 4°C. If the temperature out of the pasteurizer falls, these thermometers still indicate the higher temperature.

Immediate cooling after bottling is also a problem. Since the cap (and also any air present) has to be pasteurized, it is advisable to turn the bottles over after filling and not to start cooling until after that.

The volume of the juice decreases in the course of cooling, and a partial vacuum is created in the bottle. If the cap is not airtight or if the bottle is chipped, this partial vacuum draws air into the bottle. Microorganisms in the air come into contact with the cold juice and can spoil it. For this reason new caps should always be used and bottles should be checked before use.

Table 6.4. Pasteurization procedures in common use.

Procedure	Continuous operation	Precise temperature control	Easy cleaning	Cold-filling	Use for cloudy juices	Carbonated beverages	Output (l/h)
twin bell chamber	+	−	−	−	+	−	< 600
plate pasteurizer	+	+	+/−	−	+/−	−	> 500
tubular heat exchanger	+	+	+	−	+	−	> 500
tunnel pasteurizer	+	+/−	+	+	+	+	> 1000
chamber pasteurizer	−	+/−	+	+	+	+	> 250

+ possible, +/− depending on design, − not possible.

USE OF PROTECTIVE GASES. Apart from pasteurization, there are a few other methods of giving fruit juices a long shelf life. Covering with carbon dioxide is one of them.

This so-called Seitz-Böhi process was in great vogue in the 1930s. The juice was covered with carbon dioxide in pressurized tanks. Carbon dioxide was pumped into the tanks until the pressure was well above atmospheric, thus preventing fermentation. It was not possible to inactivate all microorganisms with the overpressures used. The fruit juices changed, and many of them suffered spoilage as a result. For this reason the Seitz-Böhi process never became very important.

THE PRESSURIZED VAT. The idea underlying the pressurized vat is based on a principle similar to the Seitz-Böhi process. Carbon dioxide produced by fermentation is supposed to build up an overpressure which prevents further fermentation in the tank. Pressure-tolerant yeast strains, however, can survive several bars of overpressure and produce much more than 0.5% by volume of alcohol before fermentation stops. Prolonged storage is only possible with clarified juices at low temperatures.

The overpressure only inhibits fermentation by yeasts, but does not inactivate enzymes and other microorganisms. These can cause changes in the colour, odour and taste of the juice.

Bottling of juices

Only stainless steel bottling equipment is in general use nowadays. This is resistant to aggressive fruit acids and is easy to clean. Pipes and tubes which come into contact with the juice must also be suitable for the bottling of hot juices.

In hot-filling, the bottling equipment should ensure that the bottles are as full as possible, and without the inclusion of air if possible. Inclusion of air in the juice results in re-infection with airborne microorganisms (mould spores, yeast cells and bacteria).

Production of dried fruit

About 80–90% of a fruit is water. The action of heat in drying evaporates the water and concentrates the remaining constituents of the fruit. The residual moisture content is critical for the keeping quality of dried fruit.

Apples, pears, plums and apricots are especially suitable for drying. With regard to grapes, only seedless cultivars are suitable. These are mainly used to produce raisins.

Pre-treatment of fruit

Mechanical/thermal pre-treatment

Fruits of approximately equal size are used for the production of dried fruit. Differences in fruit size give rise to differences in drying times. Sorting by size is essential, therefore, if a product that is uniform in appearance and degree of drying is to be obtained.

Depending on taste, the fruit can be previously peeled or crushed. The stones should always be removed. In the case of apples and pears, the core is cut out before crushing. Plums and apricots are cut in half. These fruit halves can be joined together again, but they are usually dried separately, with the cut surface upwards.

Thermal pre-treatment can be carried out before the fruits enter the dryer. This reduces both the initial microbial count of the fruit and the degree of enzymatic browning. The cells in the pieces of fruit that are heated break up, thus accelerating the subsequent drying process.

Plums have on their surface a layer of wax which slows down the evaporation of water. Heating destroys this layer, so a higher drying rate is achieved.

The most common method of thermal pre-treatment is blanching. In this process, wire baskets are used to immerse fruits or pieces of fruit into boiling water. Depending on the species and size of the fruits, the time required for blanching ranges from 10 s to a few minutes (or up to 10 min in the case of quinces).

In larger factories the fruit is steamed on the shelves of chamber dryers or on the belt of belt dryers.

Chemical pre-treatment

To prevent non-enzymatic browning, the fruit can be dipped in a 5% solution of citric acid (or in lemon juice). This is especially advisable in the case of sliced and diced apples where the acid content is low. The acid allows the light colour to be retained, while at the same time improving the taste of the pieces of apple. The colour of the fruit can be made even lighter by adding ascorbic acid to the dip.

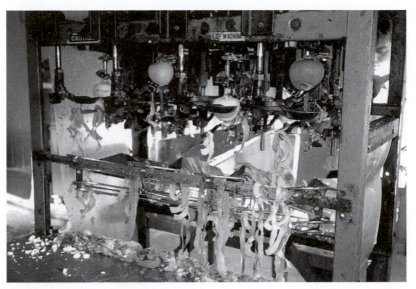

Fig. 6.7. Peeling machine in use.

Fig. 6.8. Apple slices on a stainless steel grating.

Fig. 6.9. Chamber dryer.

Suitable drying methods

Chamber or shelf dryers

The most suitable equipment for the production of dried fruit is the chamber or shelf dryer. Smaller types have strips inside, along which the drying screens (shelves) are moved like shutters. Bigger dryers are designed in such a way that a mobile stand is filled with these shelves. The stand is then pushed into the dryer. Electrical heating elements are usually employed to heat the dryers. Built-in fans provide the necessary air circulation. The better dryers are fitted with air baffle plates which distribute the warm air evenly throughout the dryer.

The drying screens or shelves consist of a frame and a stainless-steel sieve screen, or a screen with a teflon coating. Both materials act to prevent the fruit from sticking to the screen.

A distinction is made between dryers with horizontal and vertical air circulation. The first type dries the fruit more slowly but also more uniformly, and the distance between the individual shelves can be kept bigger. Dryers with vertical air circulation work faster but not as uniformly. They can have a more compact construction for a given output. During the drying process the drying screens have to be changed in order to achieve a uniform degree of drying.

The warm air is moved by a fan. The better the air circulation in the dryer, the more uniform is the degree of drying of the fruit.

Table 6.5. Other major drying methods and their characteristics.

Drying in	Continuous operation	Good control	High output	Heating
the open air*	–	–	+/–	solar energy
oven	–	–	–	electrical
desiccator	–	+/–	–	electrical
belt dryer	+	+	+	depending on design

* Widely used in southern Europe. + possible, – not possible.

Actual drying

The actual drying process takes place in two stages.

In the first stage water evaporates uniformly from the inside of the fruit to the outside through the capillaries. The temperature at the surface of the fruit and the drying rate remain to some extent constant. In this phase, which lasts until the capillaries are broken, the temperature of the surface is significantly lower than the air temperature.

If the temperature in this phase is made too high, entrained constituents block the capillaries, the surface becomes incrusted and the drying rate is greatly reduced.

In the second stage water vapour has to pass through the dry capillaries from the inside of the fruit to the outside. The drying rate diminishes, while at the same time the surface temperature of the fruit rises.

In the first phase there is still a difference in temperature between the air and the substance to be dried. The longer the drying process lasts, the warmer the fruit gets. The temperature in the second phase of drying is reduced by up to 10°C in order to avoid browning and loss of flavour.

Production of jellies and jams

Under current EU law, jellies and jams are products gelled by suitable methods from fruit with a precisely prescribed dry matter content.

Jam: the raw material is fresh or deep-frozen fruit; this is crushed and concentrated by suitable methods.

Jelly: the raw material is fruit juice; this is concentrated by suitable methods.

Keeping quality of jam and jelly

Preservation by a high dry matter content

Jam or jelly is traditionally produced by adding sugar and, if necessary, a gelling agent. The addition of sugar is important for the keeping quality. Products with a dry matter content of 60% or more are protected from spoilage by microorganisms. Certain species of microorganisms, however, can still grow at higher sugar concentrations.

Possible ways of increasing the dry matter content

ADDITION OF SUGAR.   When adding sugar it is important to make sure that it complies with the organic farming standards (organic sugar). Honey can also be used instead of sugar. Honey has between 70 and 80% dry matter.

EVAPORATION OF WATER.   Prolonged boiling of fruits increases the dry matter content by evaporating water. The fruit has to boil for hours in order to achieve a sufficiently high dry matter content. This causes distinct changes in colour and taste. The shorter the boiling time, the fruitier is the jellied product.

The dry matter content increases to some extent with every boiling.

Preservation by heat

HOT-FILLING.   Products with a dry matter content lower than 60% do not have any self-preserving effect. Such products achieve keeping quality through the boiling process and hot-filling or through strict observance of good manufacturing practice.

The heat capacity of the hot product is sufficient to kill microorganisms on the inside of the jar. Microorganisms on the lid of the jar are inactivated by heat by turning the jar over immediately after it is filled.

Jams with a low dry matter content (below 60%) do not keep as long as other jams.

PASTEURIZATION OF THE PRODUCT IN THE JAR.   A reliable method of prolonging the shelf life of jams or jellies is pasteurization of the product after the jar has been filled and sealed. Depending on the size of the jar, the product is heated in a water bath at a temperature of about 90–95°C for about 5–15 min.

This method of prolonging shelf life has not been widely adopted in practice. Hot-filling after boiling is available as part of the manufacturing process. For products with a low sugar content, pasteurization of the final product offers an additional guarantee of keeping quality.

Chemical preservatives

It is not permitted to use chemical preservatives (such as sorbic or benzoic acid or their salts) or synthetic colours in organic jams and jellies.

As a rule, the various possible ways of preserving the product are combined in manufacture. In practice, hot-filling takes place after the addition of sugar and pectin.

Gellability of fruits or fruit juice

The pectin content is critical for gellability. This varies, depending on the ripeness of the fruit used and on the species of fruit.

As a general rule, the more acid and less ripe the fruit is, the higher is the pectin content.

Table 6.6. Classification of fruits according to their pectin content.

Low	Medium	High
strawberry	blueberry	currant
raspberry	pear	gooseberry
cherry	peach	quince
apricot	elder	
plum		

Fruit of species with a high pectin content does not normally need any additional gelling agent, but a gelling agent has to be added in the production of jam from low-pectin fruit.

Gelling agents in common use

Pectin

Pectin is obtained from unripe apples or citrus fruit. In both cases the residues left after pressing to extract juice are used for pectin production. In the fruit, pectin is responsible for the firm structure. Apples, for example, have a firm texture despite a water content of more than 85%. Pectin is sold almost exclusively in powder form.

Crushed fruit or fruit juice is mixed with sugar (or honey) and pectin and boiled. In the case of products that are low in sugar it is advisable to mix the pectin with about ten times as much sugar and add it to the fruit mixture. The remainder of the sugar is then added.

Pectin can also be added in the form of a solution. The required amount of pectin is dissolved in a small amount of hot water (about 80°C) and added to the fruit mixture. In this case the pectin is not added until after the sugar. Citric acid or lemon juice is also added in order to speed up the gelling process and improve the taste. The amount of pectin to be added depends on the original pectin content of the fruit and on the size of the container. Less pectin is needed for small containers than for bigger containers.

Agar agar

This product is a highly effective gelling agent derived from algae (one teaspoonful of agar agar is sufficient to gel half a litre of liquid).

The fruit is crushed small and thoroughly mixed with the sugar (or honey). Lemon juice and agar agar are added to a small portion of the fruit purée, and this mixture is then well stirred.

In the meantime the remaining crushed fruit is heated with the sugar (or honey) and boiled for about 1 min. The gelling mixture is then stirred into the pot. Filling of jars can be started after the mixture has been brought to the boil again.

Other gelling agents

Other gelling agents sometimes have a pasty taste, so it is advisable to try them out first. They must be included on the list of permitted ingredients for organic products, however.

Gelling test

Before jams or jellies in which gelling aids (excepting agar agar) have been used are put into the jars, it is advisable to carry out a gelling test. This makes sure that the mixture really does gel when it goes cold.

The pot is usually taken off the hot-plate and a few drops of the still-boiling hot mixture are dripped on to a plate. If the mixture sets after cooling, the filling of jars is started. In order to speed up cooling, the plate can be cooled in a refrigerator beforehand.

If the drops do not set, boiling must be continued. The maximum boiling time of about 15 min should not be exceeded, otherwise the added pectin loses its gelling power.

Filling the jars

As soon as manufacture is completed and a successful gelling test has been performed, the jars that have been prepared must be filled with the hot mixture. Any foam produced during boiling is removed with a skimmer before the jars are filled.

The jars must be cleaned carefully and are filled up to the brim, if possible, and then sealed. Twist-off lids are generally used. New lids should always be used, as used lids do not usually guarantee a completely tight seal.

As soon as they have been filled, the jars are turned upside down. In this way the lid is sterilized with the hot fruit mixture. The lids can also be sterilized beforehand with alcohol or in an oven.

Products made with agar agar take up to 2 days to gel. The jars should not be moved during this time.

Changes in the jar

Sugar crystallizing out

In jams with more than 65% dry matter content, the sugar may crystallize out as a result of cold. This occurs especially when honey has been used as the sweetener.

A long boiling time, slow cooling and a high storage temperature are factors conducive to crystallization. The problem can be avoided by replacing part of the sucrose in the formulation (at least 25%) with glucose syrup.

Syneresis 'bleeding' of jams

Apart from achieving a firmer consistency, the reason for adding pectin is to bind water. If this effect is not adequately achieved, the gel contracts and juice is released in the jar. High dry-matter contents prevent contraction of the gel and thus the tendency to syneresis.

An excessively high acid content (and thus too low pH) is one of the main reasons for this fault. It can be corrected by reducing the amount of acid added.

Cold-stirred jams

Crushed fruit is mixed with organic sugar or honey that complies with organic farming standards, without heating. Approximately equal amounts of fruit and sugar are used. The target values for dry matter content are applicable here also.

Berry fruit, e.g. raspberries or strawberries, is particularly suited for the production of this type of jam.

The advantage of cold-stirred jams is that the fruit flavour is retained better than in boiled products. The disadvantage is the shorter shelf life. Cold-stirred jams cannot be kept for more than 1 month under refrigerated storage conditions. They should therefore be produced as and when required.

Cold-stirred jams can also be kept frozen. It is preferable, however, to freeze the fruit and make the jam from day-to-day in small batches.

Production of vinegar

For some years many producers have been deliberately trying to produce something that used to be regarded as an accidental and sometimes even unwanted by-product: vinegar.

There are essentially two possible ways of producing vinegar:

- **acid vinegar**: this is obtained by diluting and flavouring pure acetic acid
- **fermentation vinegar**: in this type of vinegar, acetic acid is produced by so-called acetic fermentation from alcohol. Only this type of vinegar production is permitted in organic fruit production.

As part of their metabolism, through the action of air, microorganisms convert the alcohol present into acetic acid and heat.

The amount of acetic acid produced can be determined from the chemical formula:

1 ml of alcohol yields 1.036 g of acetic acid
1 g of alcohol yields 1.3 g of acetic acid

These figures are only theoretical, however, as other substances may also be produced during acetic fermentation and these alter the figures in practice. In general terms, it can be said that 1% by volume of alcohol yields about 1% of acetic acid.

Vinegar is not completely stable until its acid content is about 5%. In order to achieve this acid content, the primary product needs to have an alcohol content of 5% by volume or more.

The acid content and clear filtration are critical for the keeping quality of vinegar.

Production factors

Primary product

The primary product, usually cider or grape wine, must have completed alcoholic fermentation and be free of fermentation residues (lees), and should not be sulphurized. A further requirement is that it should be free of odour and taste defects (in this case the product may possibly be too vinegary). With vinegar, just as in other fruit products, the quality of the primary product determines the quality of the final product.

Bacteria

The acetic acid bacteria are introduced either as 'mother of vinegar' or as a liquid culture. Small producers often use mother of vinegar. This consists of acetic acid bacteria and slime bacteria. The latter are responsible for the consistency of the mother of vinegar. They also produce metabolites, however, and for this reason acetic fermentation by mother of vinegar is not always completely pure in tone.

The second possibility is to use a liquid culture. In this case it is possible to use already fermenting vinegar or a special liquid culture. Dried formulations such as fermenting yeasts are unfortunately not yet available for vinegar production.

The production of acetic acid from alcohol is carried out by two naturally occurring groups of acetic acid bacteria:

- *Acetobacter*: these are colourless rod bacteria that live mainly on plants which secrete sap containing sugar. They are about 1/1000 mm in size and almost always live together with yeast cells.
- *Gluconobacter*: these are also rod bacteria. They exhibit high acid tolerance and are almost identical in size to bacteria of the *Acetobacter* group.

Temperature

Temperature is an important factor in vinegar production. Like all bacteria, vinegar bacteria prefer warm conditions. Temperatures around 25°C are essential for rapid and effective fermentation. If the temperature of the vinegar falls markedly during fermentation, the fermentation process is interrupted and may start again later. The temperature in the vinegar vat should therefore never drop below 25°C.

Air

Air is needed by the bacteria for their 'work'. For acetic fermentation to start, the must has to be adequately aerated. Open storage alone is often not enough.

Depending on the type and size of container there are various methods of introducing air. Small containers can be shaken, but for bigger

quantities it is advisable to use an air pump (aquarium pump or com-
pressor, with air jets in the container and a time relay).

Container

Any type of container, ranging from the wooden vat and the plastic container to the special steel tank, is more or less suitable, provided that it
is of food-grade quality and resistant to acid. If possible it should have
large apertures and drain-cocks, or facilities for controlling the admis-
sion of air.

Containers which have been used for vinegar production are no
longer suitable for the production or storage of other products. They are
usually so heavily contaminated with vinegar bacteria that other products stored in the container would inevitably acquire a vinegary flavour.
The exception is stainless-steel containers, which can be disinfected
with suitable cleaning products.

Vinegar production processes

Open fermentation process

In this process the acetic fermentation takes place in a more or less open
container (usually a vinegar vat in the cellar). The primary product stays
in the open container until acetic fermentation starts, i.e. a mother of
vinegar, or film consisting of vinegar bacteria, moulds and slime fungi, is
formed. In this method it is not possible to control the fermentation or
the temperature. The quality of the vinegar produced does not always
meet current sensory and analytical requirements.

Pumping process

The pumping process, which has been in use since about 1930, makes it
possible to produce vinegars of relatively high quality. In this process the
vinegar is circulated over shavings, maize cobs or wooden trellises. The
temperature is increased by the vinegar bacteria. The fermentation can
only be controlled by heating and cooling. With a modicum of skill it is
possible to construct the equipment for this oneself.

If there are problems with the temperature, however, the product
often has a very high aldehyde content. Fermentation takes between 6
and 10 weeks.

Submerged fermentation process

Submerged fermentation processes are currently considered to be the
best methods of vinegar production. The bacteria are active in the liquid,
no mother of vinegar is produced, and continuously controlled aeration
is guaranteed. This means that temperature control and thus carefully

planned vinegar production are possible. These processes are used mainly in large factories.

Labelling of products

All foods which are offered for sale must be labelled in accordance with the regulations in force. Only approved and inspected organic undertakings may describe their products as 'organic'. It is not permitted to use labelling elements which refer to the health of the consumer.

The labelling elements specified below must be used in connection with the products mentioned in the preceding chapters.

Reference to organic

Only officially inspected organically run undertakings may declare their products to be organic. Descriptions such as 'from organic farms' are to be used.

Since January 1997, it has been compulsory under EU Regulation 2092/91 to indicate the code number and name of the inspection body on the label.

Product name

The product name indicates what the product is.

In the case of fruit juices, the species of fruit used is indicated together with the word '... juice'. The words '100% juice' should be added to the product name.

For jam and vinegar the wording is '... jam' and '... vinegar', respectively.

In the case of dried fruit, it must be clear from the product name that the fruits have been dried.

In addition the term 'organic' can be inserted in front of the product name (e.g. organic apple juice – 100% juice).

Name and address of the producer

This information is required for all the products mentioned. The minimum requirement is that it must be possible to trace the product back to the producer. The more precise the information, the better. The code letter of the producing country should be given in front of the post code.

Batch number

A batch number is a combination of letters and numbers which can be freely chosen. The producer can choose any number desired, but when it is shown on the label it must always begin with an 'L'.

The purpose of the batch number is to make it possible to trace back any product in respect of which a complaint has been made with regard to quality, product name, etc. A lot number indicates a complete set of sales units of a food which was produced under the same conditions (same batch).

Net contents

The net contents are indicated in litres or grams. In the case of returnable bottles, this information can be omitted if it is imprinted on the lower rim of the bottle.

Minimum shelf life

A minimum shelf life of 12–18 months is usually indicated for the four products mentioned. The wording on the label is 'best before end of ... (month and year)'.

List of ingredients

All the ingredients used are indicated in decreasing order of quantity in the list of ingredients. This information must be given in letters and figures of the same size and colour. The ingredients are not indicated if no substances have been added.

Storage instructions

The storage conditions must be indicated if they are significant for the storage life of the product. The wording used as a rule is 'Keep in a cool place after opening' or 'store in a cool, dry place', or something similar.

Additional information

For some products further information must be given in addition to the labelling elements stated above.

For **jam and jelly**, respectively, information on the fruit content ('Made from ... g of fruit per 100 g') and on the dry matter content ('Total sugar content ... g/100 g') must be shown on the label.

In the case of **vinegar**, the percentage acid content ('Acid content ...%') and the kind of vinegar must also be stated. Vinegar made by acetic fermentation of an alcoholic liquid is described as 'Pure fermentation vinegar'.

7 Marketing of organic products

Selling means giving the customer something that is of benefit to him/her

Marketing is customer-oriented business management, i.e. the customer's satisfaction is the focal point of all marketing considerations.

In essence there are two possible ways of bringing one's products to the customer:

1. The producer produces and delivers to the **wholesaler**. This is the usual approach, especially in conventional agricultural production.

The second possible way of marketing is used mainly by producers of organic farm produce.

2. The producer produces and markets **direct to the consumer**:

- ex-farm
- farmers' markets
- home delivery
- delivering to restaurants and hotels
- agritourism (farm holidays).

Only surplus capacity is sold to the wholesale buyer.

For successful adoption of this direct approach to the consumer, the following points must be borne in mind, especially for organically run undertakings:

- **Environmental commitment:**
  You yourself must be convinced of the benefits of organic farming and convey this conviction to the customer.
- **Winning and keeping the customers' confidence:**
  People who buy organic products generally have a health-conscious outlook and environmental concerns and therefore want to be sure that they are really getting genuinely organic products. They also pay a higher price for them.

The seller of organic products must therefore always emphasize and demonstrate the continuous inspection of his products (control association, control number) in his marketing.

- **Product assurance:**
  The consumer wants complete assurance that the products he buys really are organic. At the same time, however, he also wants to be sure of regular delivery and constant quality.
- **Product presentation:**
  Product presentation must be in keeping with the organic production of the products: no plastics in the packaging; clear product labelling; additional information on the methods of production, etc. It should highlight the additional benefits for the customer from local production, protection of the environment and conservation of the landscape.
- **Cost awareness:**
  If you want to make a success of direct marketing, you must keep a record of your costs. This is not only for yourself, to find out whether it actually pays, but also for purposes of cost-based pricing.

The **checklists** given below will help you in your work.

Lasting success on the market is only possible if the following triangular relationship works effectively:

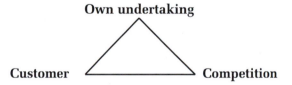

Customer

New or better products (or services) must create new benefits that are perceptible for the customer.

Competition

A competitive advantage over competitors must be achieved.

Own undertaking

- Operation must be innovative, efficient and oriented towards high quality.
- Always being ready and able to make improvements, and take pleasure in one's work.

One's own undertaking is therefore the starting point for all new thinking.

Marketing as a way of thinking and operating can be summarized in the following concepts:

1.   Analysis of the current situation:
Where are we now?
What information do we need?
2.   Goals – plans – measures:
Where do we want to get to?
What market shares? What turnover? What profit?
3.   Marketing strategy:
What path will we take to achieve our goal?
4.   Marketing instruments:
Product policy
Price policy
Distribution policy
Communications policy
Marketing mix
What marketing instruments will I use?
How can I fine-tune these marketing instruments so as to achieve harmony between them?
5.   Performance review:
Have we achieved our goals?
Where and how must we make changes?

Analysis of the situation

Which aspects should be considered at the outset?

The first thing is always the **idea**:

What services can the business provide on the market in a better, different or cheaper way than the competition because of its

- know-how
- resources
- entrepreneurial commitment
- ideas

so as to differentiate itself from its rivals?

Analysis of the initial situation

The analysis of the initial situation should help to identify the initial conditions encountered by the business in question. The aim is to pinpoint the strengths and weaknesses of the business and of the personality of the person running the business, so as to formulate future goals and alternative solutions.

The **checklists** below are designed to help you do this. Make sure you allow sufficient time for the first **analysis of the current situation**. It is impossible to make the right decisions without knowing the initial situation. The decision-making process is based on a sound understanding and detailed analysis of the initial situation. To make a good decision it is essential to have plenty of information, creativity and the courage to adopt new solutions. The steps involved in the decision-making process are outlined in the checklist entitled '*Decision-making process*'.

Checklist for analysis of the initial situation

Family situation
What do I want to achieve for myself? ....
What do I want to achieve for my family? ....
What do I want to achieve for our business? ....

Labour aspects
Do I have spare labour capacity? ....
Is reorganization likely to yield additional
    labour capacity? ....
What about the interim phase? ....

Financial and economic situation
Economic position of the business ....
    Equity capital ....
    Outside capital ....
How will finance be raised? ....

Assessment of the market and business environment
Is there a market for my product? ....
Target group ....
Price situation ....

What do I like to deal with?	**In what areas would I like to expand my knowledge and abilities further?**	
with production (business) ☐	production (business)	☐
with the market ☐	marketing	☐
with financing (capital) ☐	financing (capital)	☐
with entrepreneurial skills ☐	entrepreneurial skills	☐
In what areas do I have knowledge and experience?	**How many hours in each working week do I spend on average on:**	
production (business) ☐	production (business)	
marketing ☐	marketing	
financing (capital) ☐	financing (capital)	
entrepreneurial skills ☐	entrepreneurial skills	

Checklist for analysis of the strengths and weaknesses of the business

Situation	Strengths	Weaknesses
Products Services	e.g. organically produced	e.g. little known ....
....	....	....
Price	e.g. costed price	e.g. not acceptable to everybody ....
Distribution	e.g. direct selling	e.g. high labour cost ....
Marketing	....	....
Advertising	....	....
Image on the market	....	....

Decision-making process

What is to be done?	Who must do it?
Recognize and formulate the problem	→ Business manager
Collect data and information on the problem	→ Help through consultative service
Develop and analyse alternative solutions	→ Help through consultative service
Decide on a particular solution	→ Responsibility of the business manager
Implement the solution chosen	→ Support through consultative service
Monitor the degree of success	→ Business manager: supported by consultative service

Plans – goals – measures

There can be no general guidance for drawing up a business plan. Business plans are as different as the personalities of business people and their ideas.

It is precisely because there are many paths to success – but also to failure – that before setting up the business you should carefully review what you want to achieve and how you think you can achieve it.

The more factors included in this review and the more concrete the answers to the questions raised, the smaller will be the business risk.

The checklists below should help you identify your goals more clearly, so as to reduce the risk.

Checklist for direct sellers

yes	no	
☐	☐	Have you considered which consumers (target groups) you want your products to appeal to?
☐	☐	Do you think your products are good enough to win new customers?
☐	☐	Do you systematically make surveys of your customers' wishes and requirements?
☐	☐	Do you look at current trends in direct selling?
☐	☐	Do you respond to these trends?
☐	☐	Do you have advertising media that clearly highlight the particular advantages of your products?
☐	☐	Do you have targeted advertising of your products?
☐	☐	Do you run active sales promotions (e.g. sales letters in connection with special events)?
☐	☐	Does the business have its own prospectus?
☐	☐	Is there a business card?
☐	☐	Can the customer get comprehensive information about the business?
☐	☐	Is there specific written information for the customer about your products?
☐	☐	Is your product range differentiated from those of your competitors?
☐	☐	Are the products well (properly) presented?
☐	☐	Do they have their own labels?
☐	☐	Is there a corresponding price list?
☐	☐	Do you sell your products at a correctly costed price?
☐	☐	Is there a customer file?
☐	☐	Are there house specialities?

Checklist of questions to consider at the outset

Questions to consider	Product: e.g. organic apples	Product: e.g. dried fruit
Is there a market for my product or service?		
Development prospects	....	....
Gaps in the market	....	....
New needs	....	....
Which target group is to be addressed?		
Age	....	....
Social stratum	....	....
Income	....	....
Area of interest	....	....
Behaviour/values	....	....
How big is the market potential?		
Size of the target group	....	....
Potential financial gain per head	....	....

Continued

Checklist of questions to consider at the outset *Continued*

Which competitors are already in the market?
Estimated market share ....

Is the product/service competitive?
Quality ....
Production (processing etc.) technology ....
Price ....

How do I get customers?
Quality requirements ....
Price ....
Terms and conditions ....

What price will I charge?
Competitors' prices ....
Prices of similar products/services ....
Costing ....

How should I present my product/service?
Method of launching on the market (advertising) ....
Type of advertising strategy ....
Type of public relations work ....
Type of packaging ....

What trade channels should I choose?
Possibilities of distribution ....

What additional services should I offer?
Service, instructions for use, etc. ....

What underlying conditions must I take into account?
Natural and technical ....
Governmental and statutory ....

Do I have the necessary means of production to start off?
Capital ....
Machinery ....
Staff ....

How should my business be organized?
Consolidated ....
Corporate business ....

What staff will I employ, and how many?
Wage costs and incidental costs ....

How much capital do I need?
Equity capital ....
Outside capital ....
Grants ....

What targets should I set for myself?
Turnover ....
Quantity ....
Period up to … ....

What actions should I take to achieve my targets, and how should I measure the degree of success?

Checklist for profitability

Gross revenue
(quantity × price) ....

Variable and overhead costs
Costs of materials
(including primary products + ingredients) ....
Staff costs ....
Insurance ....
Vehicles (fuel and insurance, costs of purchase
and maintenance or mileage costs) ....
Rent, leasing ....
Postage, telephone ....
Membership fees ....
Premises costs, energy, heating ....
Cleaning ....
Consultancy costs ....
Maintenance and repairs ....
Tax ....
Advertising ....
Packaging ....
Other costs ....

Total

Gross revenue ....
− variable costs ....
= CONTRIBUTION MARGIN

Investment costs
Building work ....
Fixtures and fittings ....
Machinery and equipment ....
Market stall ....
Vehicles ....
....

Total investment costs

Calculation of the capital costs of the
investment (net present value method)
average useful economic life 10 years
building work 15–20 years
CAPITAL COSTS
(investment costs × annuity factor) ....

CONTRIBUTION MARGIN ....

− TOTAL CAPITAL COSTS
   (depreciation and interest) ....

= PROFITABILITY
PROFITABILITY per month

Care should be taken in deciding how much of the investment costs should be financed
with equity capital and how much with outside capital.

How big is the market?

1. **Where do I want to sell?** ....
   Define the region (this is critical for distribution)

2. **Who do I want to sell to?** ....
   Decide on target group, see *'Finding the right target groups'*

3. **How big is the sales volume?** ....
   Per capita consumption and food expenditure statistics are useful

4. **Market share** ....
   The quantity of these products sold by the biggest enterprise on the market (statistics can be obtained from chambers of commerce or statistics departments). Once some idea has been obtained of the size of the market, it will be possible to make a better assessment of one's own market prospects, determine one's sales targets and look out for any niche opportunities.

Finding the right target groups

Who do my products (or services) appeal to?	How do I want to sell?
age	direct from the farm ....
sex	delivery ....
education	door-to-door sales ....
income	farmers' markets ....
profession	my own shop in the town ....
household size	hotels ....
place of residence	bulk buyers, etc. ....
size of place of residence	

Patterns of behaviour

What is the communication behaviour like?

What media are used, how do I reach the target group with particular media?

Buying behaviour	when do people shop (time)?
	where?
	how long for?
Leisure behaviour	sport
	culture
	family-oriented
Information	(how can I give people information?)
	farm tours
	farm fêtes
	leaflets
	mail shots, etc.

One point where the marketing of organic products differs from conventional marketing is the choice of the target group.

Special target groups for the sale of organic products are people who:

- have high health awareness
- attach great importance to environmentally acceptable production
- have a high level of confidence in organic farming
- are willing and able to pay the somewhat higher price for organic products
- feel concerned about environmental issues.

How do I approach the customer?

When you have found the target group for your product, you then have to sell the products. It should be borne in mind that selling and buying are governed by a particular process.

The art of selling revolves around converting the interested (potential) customer into a purchaser. It is helpful if you as the producer (seller) can put yourself into the situation (role) of the purchaser. Even if you do not sell direct to the consumer but only deliver to a bulk buyer, you should still have a clear picture of the buying process and be able to judge sales successes and failures according to the inherent laws of the buying process.

Customer behaviour is essentially dependent on observance of the laws of the buying process. If you ignore customers and do not provide them with information, you will not persuade them to buy, even though they may be interested in the product itself.

Marketing strategy

What are the possible approaches for achieving success on the market?

Just as there are many different types of business, there is a wide variety of possible strategies for success. For this reason just two examples of possible approaches will be given here.

Expansion of the business – growth strategy

Market penetration

The products remain essentially the same. The aim, however, is to win new customers who are similar to the existing ones (income, interests, family status, etc.), while at the same time taking even greater care of the existing customers.

Overall assessment: high probability of success at little cost (compared with other possible approaches).

Example: a grower engaging in direct marketing of organic fruit products offers his products for sale at a farmers' market in a town where he has a good reputation because of the range and quality of his products.

So as to be able to meet demand even better, he decides to send his produce to two more farmers' markets in the same town.

Market development

The products remain essentially unchanged. Now, however, new markets are opened up and new customers won – either completely new categories of buyers or new sales areas.

Overall assessment: the likelihood of success is still high, but not as high as with the market penetration approach. At the same time there is also an increase in costs. These are significantly higher than in the product development and market penetration approaches.

Example: the grower selling organic fruit decides not just to offer his products for sale at farmers' markets but also to supply higher-education establishments, so as to obtain access to new markets and new customers.

Product development

New products also create growth. They can be completely new products (product innovation) or further developments of existing products. Existing customers continue to be supplied, but with new products.

Overall assessment: the risk of things going wrong is relatively high. The cost of this approach is higher than for the market penetration approach, although lower than for the other approaches.

Example: the grower engaged in direct marketing makes dried fruit products from organic apples.

Diversification

New consumers are sought for new products. The existing customers are thus abandoned.

Overall assessment: high risk and high costs. This approach is suitable for large concerns and/or mergers.

Example: the organic fruit grower decides to offer fresh meat for sale in future as well as his existing products.

How the buying process interacts with the selling process

The buying process	Organization of the selling process
Attentive and considerate approach	product choice general presentation packaging product tasting
Information	advertising and public information events production of media, etc.
Acceptance for oneself Sympathy Willingness to buy	image, self-awareness corporate identity values in vogue (e.g. environment) motivation, conviction
Decision to buy	authentic contact point (straightforward procedure) the customer must be talking to the right person prompt service formal correctness adapt to modern management systems (e.g. PC, fax)
Purchasing behaviour	reliable transport organization vehicle driver not constantly changing timing – at the right place at the right time courteous handling of complaints, if 'something happens'
Delivery – payment	sensible price policy (reliability) rational forms of payment everything must be just right
Use – consumption	advice on storage and suggested recipes re-ordering, delivery dates
Satisfaction Recommendatory behaviour	loyalty bonus for regular customers special farm tours invitation to farm fêtes

Sales promotion measures – sales strategy

Discount strategy

Sales are based on low price. All possible ways of achieving cost advantages must therefore be utilized: low production costs, high numbers of units, cost-effective distribution, low staff and administration costs, high cost awareness.

Overall assessment: this approach is very dangerous for small and medium-sized businesses, because adequate numbers of units can only be achieved through high investment costs.

There is too much competition from supermarkets.

Preference strategy

Here the decision to buy is based not on price but on a particular quality characteristic. This may be the quality of the product itself or an additional feature relating to the delivery of the product, the method of production (e.g. organic), pick-your-own, etc.

Overall assessment: this approach is usually the best one for small and medium-sized businesses.

What is important in both approaches is that the customer must know what your product stands for. The benefit to the customer must be clearly discernible.

Marketing instruments

Product

The basic prerequisite for good sales is indubitable product quality. Organic fruit meets all the requirements in this respect. In addition, the more innovative a product is (e.g. a new variety of fruit, new processed product, new type of flavour), the more attention it will attract on the market and among customers.

This is why it is very important always to keep abreast of the latest developments with regard to choice of cultivars, the best production and plant protection techniques, and possibilities for fruit processing.

You should therefore ask yourself the following questions from time to time:

- Is my quality right?
- Is my product range right?
- Is my service right (e.g. delivery, opening times)?

Complete confidence in the product

In the case of organic products, in particular, it is very important for the customer to have complete confidence in the product. The customer must be absolutely certain that what he is buying really is organically grown fruit.

The consumer gets this absolute certainty from the inspections to which organic growers are subject and which are documented by an

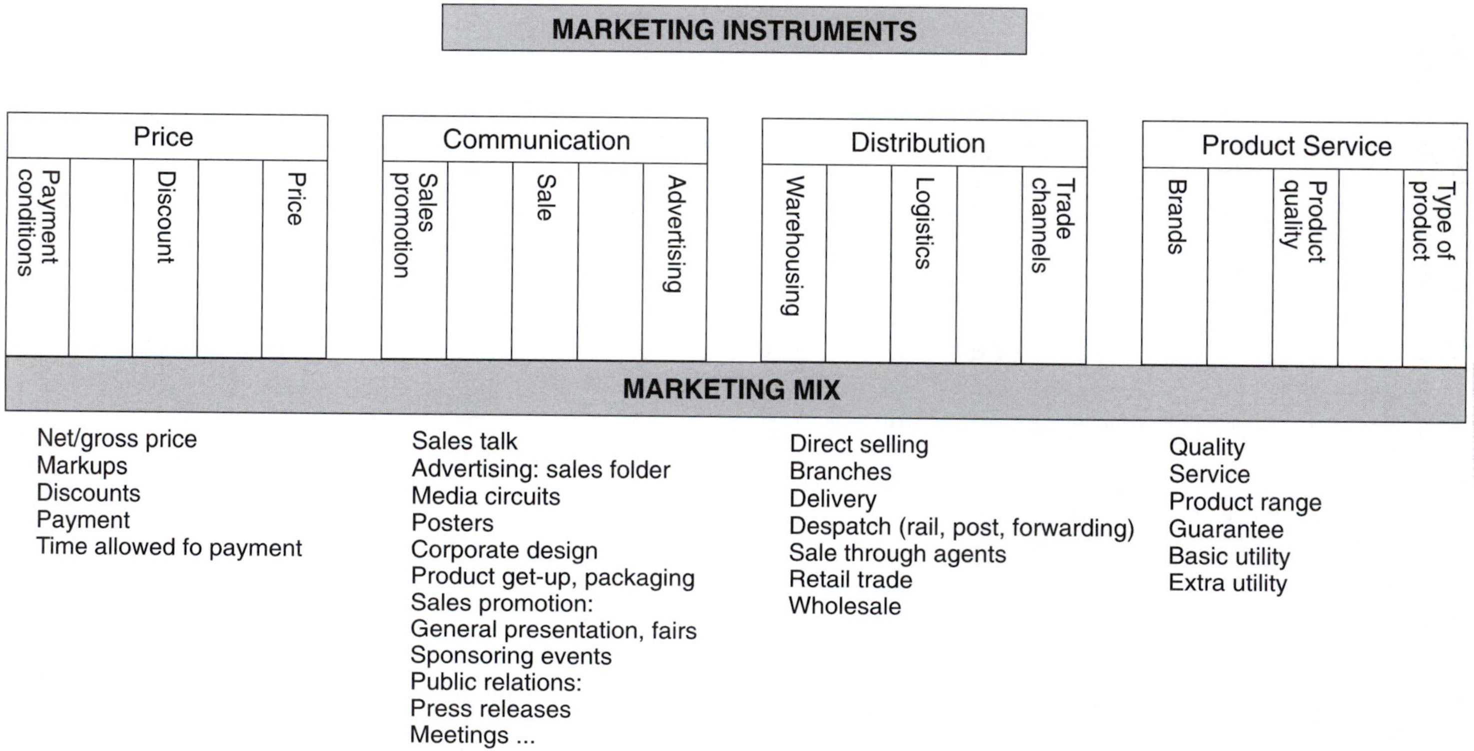

MARKETING INSTRUMENTS

MARKETING MIX

Product Service			Distribution			Communication			Price		
Type of product	Product quality	Brands	Trade channels	Logistics	Warehousing	Advertising	Sale	Sales promotion	Price	Discount	Payment conditions

Product Service:
Quality
Service
Product range
Guarantee
Basic utility
Extra utility

Distribution:
Direct selling
Branches
Delivery
Despatch (rail, post, forwarding)
Sale through agents
Retail trade
Wholesale

Communication:
Sales talk
Advertising: sales folder
Media circuits
Posters
Corporate design
Product get-up, packaging
Sales promotion:
General presentation, fairs
Sponsoring events
Public relations:
Press releases
Meetings ...

Price:
Net/gross price
Markups
Discounts
Payment
Time allowed fo payment

Fig. 7.1. Marketing instruments and marketing mix.

inspection number. You should inform your customers about the production standards with which you have to comply as an organic grower, and which provide a guarantee for the consumer.

Utility for the customer

Never forget to make sure that the buyer clearly understands the utility of your product. The customer must always have the feeling that the utility justifies the costs.

Every product and every service has a basic utility and an extra utility, a tangible utility and an intangible utility.

Every customer buys both with his 'head' (basic utility and tangible utility) and with his 'stomach' (extra utility and intangible utility).

The customer does not just buy your product, but also your entire way of life as a fruit grower.

Just think, for example: why should anyone buy a dried fruit bar made from organic apples?

What is the basic utility?	→	appeasing hunger (snack between meals)
What is the extra utility?	→	health (e.g. better digestion, lowering cholesterol)
What is the tangible utility?	→	natural product, no chemicals
What is the intangible utility?	→	improvement of the environment, conservation of landscape, way of life as a fruit grower, easing a guilty conscience, etc.

What are the basic utility and extra utility of your product?

Successful businesses do not just sell a product, but a feeling – a message!

Distribution

In what way does the marketing of organic products differ from conventional marketing?

The most typical feature hitherto has probably been that each producer of organic products has tried to achieve successful sales as an individual business.

In so doing, producers have tended (and still tend) to overlook the fact that sales are based on chance, but marketing goes through a series of stages:

- **Stage 1**
  The producer of organic products starts direct marketing and takes pleasure in every single sale.
- **Stage 2**
  The direct seller now already has a small number of regular customers; sales revenue increases relative to the first stage.
- **Stage 3**
  Success in terms of sales has now reached a peak because of the

increased number of regular customers, but unfortunately labour capacity has also reached a limit.

- **Stage 4**
  Because of the limited time and staff resources, the sales success cannot be increased any more and thus stagnates.
- **Stage 5**
  In this phase the direct seller has to make a decision, as follows. Can he carry on with the successful sales he has achieved by direct marketing, or does he need to sell the bulk of his produce via large-scale marketing structures (supermarkets)?

  The direct seller thus has to decide whether to give up direct marketing and sell to a bulk buyer or cooperate with other direct sellers and engage in joint marketing with them, taking into account the inherent laws of the marketing process.

  It is of course also possible to adopt a mixed approach, selling part of the produce directly and marketing the remainder via large-scale structures.

In essence, the following possible marketing approaches are available to the producer of organic fruit and processed fruit products:

- **direct sale to the consumer**:
  from the grower's own establishment
  farmers' markets
  home delivery
  delivery to restaurants and hotels
- **cooperation (merger) between direct sellers**
- **sale to bulk buyers**.

Decide on the method of sale at the outset

To save yourself time and money and to avoid disappointment, before starting organic production you should consider how you want to put your organically grown apples on to the market. This should also be taken into account in the choice of cultivar. Cultivars are still chosen far too often not because they sell well but because they are easy to grow.

Continuous delivery

The customer wants to receive deliveries from his suppliers throughout the year and not just for a few months. Constant liaison with customers can be built up and maintained only through continuous delivery, service and customer support (e.g. communication with the customer).

This certainty of supply is important to the customer and must therefore form an integral part of the seller's plans. That means that the seller should not only think in terms of his own particular business but should also try to find partners with whom he can work. Working together not only provides greater strength but also gives the customer the required certainty of supply. This should be demonstrated to the outside world in the form of one's own logo, packaging and presentation.

Communication

Communication does not merely mean a farm prospectus and business card. Communication is constant contact with the customer, using all available marketing instruments, and constant communication in the areas of production, plant protection, fruit processing and marketing, so as to keep abreast of the latest information.

Communication is a process that must take place in both an inwards and outwards direction. Anyone who wants to sell organic products must convey their own environmental commitment to the consumer.

Examples of ways of doing this are farm tours, demonstration gardens, demonstration distilleries, open days, 'fantasy' orchards, etc. This environmental approach should also be expressed in other areas, such as packaging, transport, choice of advertising materials and even the introduction of new products in the fruit processing sector.

The biggest marketing asset of the organic grower in particular is the trust of the customer.

There are good reasons why this customer trust is especially important for establishments that sell organic products, since the customer sometimes has to pay a higher price for the product and may have to travel further to buy it. Moreover, as a general rule, the consumer cannot see that organically grown fruit really is organically grown fruit. The consumer must therefore believe and trust the person selling organic produce.

This trust should therefore be built up, fostered and deepened by constant communication with the customer, so as to give the customer the necessary confidence in the product (brand philosophy).

Price

The purchaser expects a balanced **price–performance ratio**. This arises both from the costs involved in producing the product and from the **value**, quality, performance, shelf life and prestige. In other words, the extra utility (**value added**) that a product must have.

Competitors' pricing must of course also be taken into account.

In the case of organic products in particular this extra utility (value added in terms of health and production, that conserves the landscape and does not harm the environment) can very easily be conveyed to the customer.

The price is a figure which is derived both from cost-based pricing and from customer acceptance. One example of this is customer acceptance of branded products. Although brands are sold at significantly higher prices than no-name products, these high prices are accepted by customers (customer confidence, prestige, etc.).

When is a higher price accepted?

- When the product is needed urgently
- when it is a rare product
- when a symbolic value is offered which cannot be attained by any other product (e.g. prestige, assuaging a guilty conscience, care for the environment, shopping experience).

How cost-based pricing works

Production costs (e.g. organic apples)   ....
(fixed costs, variable costs, imputed costs)
Marketing costs   ....
(fixed costs, variable costs, imputed costs)
= Producer price   ....
+ markup   ....
= Net selling price   ....
+ taxes   ....
= Gross selling price   ....

Efficiency review

Check the goals that you have set yourself:

- turnover
- profit
- market acceptance
- communication media
- customer satisfaction
- defined goals and objectives, etc.

The following profit calculation table is designed to help you in this review.

Calculation of the profit from direct marketing

Revenue from direct marketing
(quantities sold × price) ....

minus costs of direct marketing ....

variable costs: ....
raw material costs ....
operating costs ....
outside wages ....
marketing costs ....
costs of supplies ....
....
....
....

fixed costs: ....
depreciation ....
maintenance ....
interest ....
other fixed overheads ....
....
....

= revenue from direct marketing
(profit before tax) ....

A **customer survey using a questionnaire** can also help you to monitor customer satisfaction and your communication performance and may at the same time suggest new ideas.

Appendix

Annex II to EU Regulation 2092/91 (last amended by EU Regulation 1488/97)

A. Permitted fertilizers and soil conditioners

General conditions for all the products

- Use only in accordance with the provisions of Annex I.
- Use only in accordance with the provisions of the legislation applicable to fertilizers in the relevant Member State.

Organic fertilizers

- Farmyard manure (including dried and/or composted manure) – only from extensive husbandry[1]
- liquid animal excrements (slurry, urine, etc.)
- composted household waste (source separated waste, tested for heavy metals)
- peat (only for horticultural purposes)
- mushroom culture wastes
- dejecta of worms (vermicompost) and insects
- guano
- composted mixture of vegetable matter
- products of animal origin: blood meal, hoof meal, horn meal, bone meal (or degelatinized bone meal), fish meal, meat meal, feather, hair and 'chiquette' meal, wool, wool waste from felt production, fur, hair and bristles, dairy products
- products and by-products of plant origin for fertilizers (oilseed cake

[1] Extensive husbandry in the sense of Article 6 (4) of *Council Regulation (EEC) No. 2328/91*, as last amended by *Regulation (EC) No. 3669/93*.

meal, cocoa husks, malt culms, draff, molasses, vinasse, etc.) – also 'Biosol'
- stillage and stillage extract
- seaweeds and seaweed products
- sawdust and wood chips (from wood not chemically treated after felling)
- composted bark.

Mineral fertilizers

Clays

- Bentonite, kaolin
- perlite, vermiculite.

Phosphorus

- Soft ground rock phosphate; hyperphosphate
- aluminium calcium phosphate (only if pH > 7.5)
- Thomas phosphate.

Potassium

- Potassium sulphate containing magnesium salt.

Calcium

- Calcium carbonate of natural origin: chalk, ground limestone, phosphate chalk
- industrial lime from sugar production
- calcium and magnesium carbonate: magnesian chalk, ground magnesium limestone, magnesite
- calcium sulphate (gypsum)
- calcium chloride solution: for treatment of apple trees after identification of calcium deficit.

Magnesium

- Magnesium sulphate: kieserite, Epsom salts.

Other mineral fertilizers

- Elemental sulphur
- trace compounds: B, Zn, Fe, Cu, Mn, Mo, … sulphates or chelates
- sodium chloride: rock salt (questionable, as it contains chloride)
- stone meals.

Composite mineral fertilizers

- Wood ash (from wood not chemically treated after felling).

The following are definitely NOT PERMITTED:

- Mineral N fertilizers
- readily soluble phosphorus fertilizer (superphosphate)
- fertilizers containing chloride
- quicklime, hydrated lime
- sewage sludge and sewage sludge compost.

B. Products for plant protection

The following products for plant protection are permitted for use in organic fruit growing in EU Regulation 2092/91:

Name	Description, compositional requirements, conditions for use
Azadirachtin extracted from *Azadirachta indicta* (neem tree)	**Insecticide** can only be used on mother plants for the production of seeds and on parent plants for the production of other planting stock can be used on ornamental plants
Pyrethrins extracted from *Chrysanthemum cinerariaefolium*	**Insecticide**
Quassia extracted from *Quassia amare*	**Insecticide, repellent**
Rotenone extracted from *Derris* spp. and also *Lonchocarpus* spp. and *Terphrosia* spp.	**Insecticide** need recognized by the inspection body or inspection authority
Microorganisms (bacteria, viruses and fungi), e.g. *Bacillus thuringiensis,* granulosis virus, etc.	Only products, not genetically modified, as defined in Council Directive 90/220/EEC
Pheromones	**Insecticide, attractant** in traps and dispensers
Metaldehyde	**Molluscicide** only in traps containing a repellent to higher animal species only during a period expiring on 31 March 2002
Copper in the form of copper hydroxide, copper oxychloride, (tribasic) copper sulphate, cuprous oxide	**Fungicide** only during a period expiring on 31 March 2002 need recognized by the inspection body or inspection authority
Fatty acid potassium salt (soft soap)	**Insecticide**
Lime sulphur (calcium polysulphide)	**Fungicide, insecticide, acaricide** only for winter spraying of fruit trees, olive trees and vines

Continued

Name	Description, compositional requirements, conditions for use
Mineral oils	**Insecticide, fungicide** only on fruit trees, vines, olive trees and tropical plants only during a period expiring on 31 March 2002 need recognized by the inspection body or inspection authority
Sulphur	**Fungicide, insecticide, repellent**

Index

Figures in **bold** indicate major references. Figures in *italic* refer to diagrams, photographs and tables.